THE MANAGEMENT

British Association for the Advancement of Science books published by Macmillan

SERIES EDITOR: David Reisman

Douglas Hague (*editor*) THE MANAGEMENT OF SCIENCE
Phyllis Deane (*editor*) FRONTIERS OF ECONOMIC RESEARCH
Aubrey Silberston (*editor*) TECHNOLOGY AND ECONOMIC PROGRESS
Alan Williams (*editor*) HEALTH AND ECONOMICS
R. D. Collison Black (*editor*) IDEAS IN ECONOMICS
Kenneth Boulding (*editor*) THE ECONOMICS OF HUMAN BETTERMENT
Roy Jenkins (*editor*) BRITAIN AND THE EEC
R. C. O. Matthews (*editor*) ECONOMY AND DEMOCRACY
Jack Wiseman (*editor*) BEYOND POSITIVE ECONOMICS?
Lord Roll of Ipsden (*editor*) THE MIXED ECONOMY

THE MANAGEMENT OF SCIENCE

Proceedings of Section F (Economics) of the British Association for the Advancement of Science, Sheffield, 1989

Edited by Douglas Hague

Associate Fellow, Templeton College, Oxford

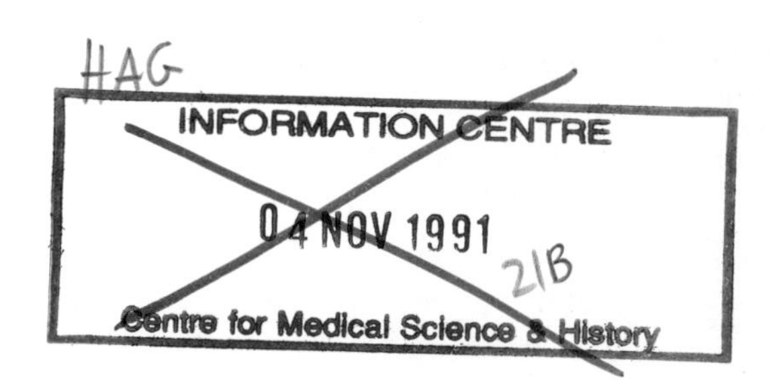

MACMILLAN

First published 1991

Published by
MACMILLAN ACADEMIC AND PROFESSIONAL LTD
Houndmills, Basingstoke, Hampshire RG21 2XS
and London
Companies and representatives
throughout the world

Printed in Hong Kong

British Library Cataloguing in Publication Data
British Association for the Advancement of Science, *Section F(Economics)*
The Management of Science.
1. Science
I. Title II. Hague, D.C. (Douglas Chalmers), *1926–* III. Series
500
ISBN 0–333–52539–6 (hardcover)
ISBN 0–333–52540–X (paperback)

Contents

Acknowledgement vi

Notes on the Contributors vii

Introduction by Douglas Hague ix

1 Can Scientists Manage Science? *Douglas Hague* 1

2 What Do We Know about the Usefulness of Science? The Case for Diversity *Keith Pavitt* 21

3 Are Some Science Policy Issues Inevitable, Irresolvable and Permanent? *Frederick Dainton* 47

4 The Management of Science in the 1990s: An American Perspective *Roberta Balstad Miller* 69

5 The Management of Pure and Applied Science Research in Academia *Eric Ash* 85

6 Finance Policy and High Politics in a European Scientific Laboratory: The Conflicts over Financing CERN in the Late 1950s and Early 1960s *John Krige* 98

7 European Countries in Science-based Competition: The Case of Biotechnology *Margaret Sharp* 112

8 Overseas Funding for Industrial R&D in the United Kingdom *Paul Stoneman* 131

9 Public Understanding and the Management of Science *Brian Wynne* 143

Index 171

Acknowledgement

I should like to acknowledge the contribution to the preparation of this book of Dr D. Reisman, editor of the Section F Series, and his colleagues on the Section F Committee. I should also like to thank the authors of the papers for the thought, time and trouble they took in preparing and presenting them.

DOUGLAS HAGUE

Notes on the Contributors

Eric Ash has been Rector of Imperial College, London, since 1985. His numerous publications include papers on patents, physical electronics and selected areas of engineering. He was Professor of Engineering at University College, London from 1967 to 1985.

Frederick Dainton is a physical chemist who has been a lecturer at Cambridge and a Professor at Leeds and Oxford and whose researches have been concerned with problems in reactions, kinetics and mechanisms, especially in chain reactions, polymer-, radiation-, and photo-chemistry. He has also been Vice-Chancellor of Nottingham University, Chairman of the Council for Scientific Policy, The Advisory Board for the Research Councils, The University Grants Committee, the British Library Board and the National Radiological Protection Board. In nominal retirement he is Chancellor of Sheffield University, Chairman of the Council of the Royal Postgraduate Medical School and a Governor of the London School of Economics.

Douglas Hague is Chairman of the Metapraxis Group of companies, an Associate Fellow of Templeton College, Oxford, and Honorary Visiting Professor at both Manchester Business School and the Management School, Imperial College, London. He has considerable experience of government bodies, having been Deputy Chairman of the Price Commission in the 1970s and Chairman of the Economic and Social Research Council from 1983 to 1987. He was a personal economic adviser to Mrs Thatcher from 1966 to 1979. Sir Douglas's writing covers a considerable field, from *A Textbook of Economic Theory* (written with A. W. Stonier) to interdisciplinary work on the role of quangoes in British and American Government. He also wrote (with Geoffrey Wilkinson) *The Official History of the Industrial Reorganisation Corporation.*

John Krige is currently the leader of the History of CERN project which is based in Geneva and independently financed by contributions from a number of CERN member states, including the United Kingdom. He was a major contributor to the two-volume *History of CERN* and is the author of a large number of articles in both natural and social sciences. He is also a part-time Visiting Fellow of the Science Policy Research Unit, University of Sussex.

Roberta Balstad Miller is the Director of the Division of Social and Economic Science at the National Science Foundation in Washington, DC. She is the author of *City and Hinterland: A Case Study of Urban Growth and Regional Development* and co-editor of *Science Indicators: Implications for Research and Policy.*

Keith Pavitt is Professor of Science and Technology Policy at Sussex University and Deputy Director of the Science Policy Research Unit. He is joint author of *The Economics of Technical Change and International Trade* and has published numerous papers on the economics and management of science and technology.

Margaret Sharp is a Senior Research Fellow at the Science Policy Research Unit, University of Sussex. She has written extensively on Europe and new technologies, including biotechnology. Her books include *The New Biotechnology: European Governments in Search of a Strategy* and *Strategies for New Technologies.*

Paul Stoneman is Reader in Economics at the University of Warwick. Dr Stoneman is the author of *The Economic Analysis of Technology Change, The Economic Analysis of Technology Policy* and many other books and articles on the economic aspects of technology and technology-related issues.

Brian Wynne is Reader in Science Studies and Director of the Centre for Science Studies and Science Policy at Lancaster University. Since the 1970s he has taught and researched on the sociology of science, focusing especially on the authority of scientific knowledge in public issues such as risk management. He is the author of *Rationality and Ritual: The Windscale Inquiry and Nuclear Decisions in Britain* and *Risk Management and Hazardous Wastes: Implementation and the Dialectics of Credibility* and co-editor (with Roger Smith) of *Expert Evidence: Interpreting Science in the Law.*

Introduction

DOUGLAS HAGUE

This book contains the papers presented to the Economics Section (Section F) of the British Association during its annual meeting in Sheffield in September 1989. The theme–The Management of Science–was chosen for two main reasons. First, science policy is a relatively new field of study to which an inevitably small group of specialists are making increasingly important contributions. More than half the papers in this volume are contributed by such specialists, namely, Keith Pavitt, John Krige, Margaret Sharp, Paul Stoneman and Brian Wynne.

Second, the issues which science policy tackles are themselves becoming increasingly important. In the United Kingdom, if one aggregates expenditure on research in science and technology in universities, the government is at present providing around £1 billion per annum of finance. Part of this is funded through the University Finance Committee and the remainder through the government's Research Councils. Even £1 billion may be a smallish sum in terms of total government spending or total gross domestic product (GDP) in the United Kingdom, but it is nevertheless a very considerable sum of money, though one which most scientists would like to see substantially increased.

It is generally accepted that the current phase of development in all industrial countries is strongly based on scientific research and development. It is therefore important that expenditure on research in the United Kingdom should be large enough–but not larger than is needed–to allow Britain to benefit as fully as possible from research in science and technology. These are crucial issues.

My own chapter sets the scene by considering the contribution which economics and the other social sciences can make to science policy research, not least by asking the right questions. These are questions which, because of their training, scientists rarely ask themselves. The chapter puts these questions in a practical context, relating them especially to the way in which the scientists, businessmen and civil servants who are responsible for managing science policy at the highest level in the United Kingdom tackle the issues which confront them. It concludes that science policy will be most effective if made by interdisciplinary teams, where social science–if not social scientists–is adequately represented.

One of the dangers of debates on science policy between scientists and non-scientists is that they fluctuate between grand philosophical arguments and generalisations, often of dubious value, based on particular examples. As an antidote to this, Professor Pavitt was asked to provide an analysis of what is actually going on. A distinguished member of the Science Policy Research Unit (SPRU) at the University of Sussex, Professor Pavitt shows that, as usual, the conventional wisdom is not always correct. For example, he questions the belief–which indeed I quote in my chapter–that science is now in a steady state, showing that there is not stagnation but vigorous growth in the numbers of scientists and engineers employed in both the United States and Japan. The United Kingdom does, however, seem to be lagging, with the UK public sector lagging most of all. Since that was the main concern of my own chapter, perhaps I am right to accept the 'steady state' there. We must remember, however, the precept of general systems theory that a large system will always be more stable than some of its sub-systems.

Professor Pavitt also shows that the impact of science on technology is more complex than most public discussion suggests and goes on to question the belief that science-based research is increasingly feeding directly into technology. Not least important, he challenges the belief–especially in some quarters in the United Kingdom–that research will be more effective if concentrated in fewer institutions.

On this basis, the discussion is taken in several directions. First, as a counter-balance to my own views, Lord Dainton–a distinguished scientist and scientific administrator–looks at similar issues to mine, but as a scientist. Drawing on his enormous experience, Lord Dainton's chapter gives an authoritative survey of the key issues in science policy today in the United Kingdom. I find it interesting that the only substantial difference between Lord Dainton and me is over the efficacy of management techniques like corporate planning, though I would not retract one word from what I say on this in my own chapter. There is much more common ground between us than is often supposed by the critics and practitioners of science in the United Kingdom.

We must, however, remember how parochial our debates in Britain so often are. The second shift in direction therefore takes us to an American perspective, in Dr Miller's chapter. As frequently happens, experience in the United States sharpens our perceptions of developments in Britain and, indeed, changes them. Dr Miller suggests that 'the informal contract between science and society . . . during World War II has dissolved'. We might not put this quite the same way in the United Kingdom, but is not something similar happening here? In this and other ways, Dr Miller's chapter invites us to take a different view of British science, because she looks at it through American eyes.

Thus far, the chapters have dealt with science policy at the national and international level, but scientific work is actually carried out in institutions. Many of these are universities. Sir Eric Ash's chapter gives a sensitive and perceptive analysis of the issues which are of greatest moment to those responsible for managing science in an academic institution–in this case his own, Imperial College, London.

It would, again, be only too easy to take a parochial British view of institutions. Dr Krige's chapter therefore takes us outside the United Kingdom to a study of CERN, the nuclear research organisation based in Switzerland. He brings not merely a European but an international dimension. Certainly the chapter reveals some of the problems of managing an organisation which has a budget of about £325 million – several times that of Imperial College. It also goes well beyond this, to consider the issues of high finance and high politics which arose during the first two decades of CERN's existence. With the rising costs of science around the world – not least for instrumentation and experimentation – revealed by the other chapters, international research organisations like CERN will become increasingly common as we move into the twenty-first century. The lessons to which Dr Krige points can therefore help us to make fewer mistakes in setting up such organisations in the future.

Thus far, the book has concentrated on scientific research funded by the public sector but, in the current phase of the industrial revolution, private-sector businesses are playing a leading role through the developments they are bringing about in high technology and information technology. A typical field is biotechnology. Margaret Sharp's chapter studies the roles of small and large firms and of both the public and private sector in the development of biotechnology; indeed she throws welcome light on some of the complementarities between these. In the process she adds to our understanding of the part played by venture capital in financing biotechnology and, especially, of the difference between its role in the United States and in Europe.

Margaret Sharp, more than any of the other authors, considers how academic science and business work – or do not work – together, by studying how they are pushing forward the frontiers of research in this exciting field. The fact that few researchers have looked, as she does, at the strengths and the frailties of the links between the academic and business communities makes Margaret Sharp's contribution especially valuable.

Another characteristic of the international financing of science today is the amount of expenditure by companies in one country on research and development in another. Dr Stoneman concentrates on the funding by overseas private companies of industrial research and development in the United Kingdom. He shows that such expenditure is not only substantial and growing, but that it appears to be greater than in comparable countries and that it is concentrated in four main industrial sectors. On this basis, Dr Stoneman points to unresolved issues and policy questions which could be illuminated by further research.

Ten years ago – perhaps even five years ago – a volume like this could have ended there. Today it cannot. There is now more interest, indeed concern, over what has come to be called 'the public understanding of science'. If this concern is new, so is its study. Dr Wynne is one of the pioneers of such research. His chapter is important because it is an early venture into a field of research which will grow, now that 'green' issues are attracting greater public interest. His study is also important because it is based on public reactions to an event which, perhaps more than any other, has accentuated concern over the effects of scientific development – the contamination of Cumbrian sheep farms by fall-out from the nuclear disaster at Chernobyl.

This book reviews some of the liveliest current issues in science policy in the United Kingdom and puts them in an international context. In its course, the book points to further issues which merit informed research and public discussion.

1 Can Scientists Manage Science?

DOUGLAS HAGUE

INTRODUCTION

Organisations throughout the public sector in Britain – not least the universities, the government-funded Research Councils and publicly-funded laboratories – are being compelled by government pressure in the form of financial stringency to take management more seriously than ever before.

For as far ahead as any of us can see – and whatever government is in power – funds for sciencc will be constrained. As John Ziman, Chairman of the Science Policy Support Group has said, science is now 'moving into a dynamic steady state in the sense that adjustments to change have to take place within a roughly constant envelope of resources'.[1] To get best value from the resources devoted to science, something has to change; a different approach is needed.

Scientists find this insistence on the need for changed behaviour disconcerting because it comes at the end of a long period during which expenditure on science grew rapidly. In part, this resulted from the belief that, since we live in an age when economic development is strongly based on the exploitation of science, greater expenditure on science will therefore always be beneficial to the nation. And it will be beneficial whether or not the discoveries arising out of that expenditure are being exploited by the rest of the community or, indeed, could be.

There is at least the implication here that more science is always better for the nation than less, so that we should be unstinting in its support. That is what led Sir Peter Middleton, Permanent Secretary to the UK Treasury, to suggest that Britain has created a 'science mountain'. Though they hate to be told so, scientists are being selfish with society's resources. Like many artists, scientists think they have a divine right to appropriate society's resources for themselves, simply becausc of their calling. Society has the right to impose on scientists the responsibility of being accountable to it for their expenditures – and their actions.

For most of my career, I would have treated Peter Middleton's remark with wry amusement. However, when I became Chairman of the Economic and Social Research Council (ESRC) in September 1983, I discovered that this gave me automatic membership of the Advisory Board for the Research Councils (ABRC). This Board has the responsibility for advising the Secretary of State for Education and Science on what should be the total amount of money allocated to what is called the Science Vote – most of which is shared between the Research Councils. There are five of these, of which the Economic and Social Research Council is one. The Science and Engineering Research Council is much the biggest of the remaining four, the others being the Medical Research Council, the Natural Environment Research Council and the Agriculture and Food Research Council. The ABRC also advises on the allocation of funds between the five councils.

Inevitably, when I joined the ABRC, I looked at the way it worked through the eyes of someone who began life as an economist but who became involved in management education in 1957 and has not left it. One thing struck me forcibly and immediately. The ABRC was concerned with managerial questions, yet no one saw them as such. Instead, it seemed a quixotic body. The other members of the ABRC were on a different wavelength from me. This is part of the British disease: it is not simply a question of there being two cultures; as members of those cultures, we transmit on different wavelengths and so do not communicate with each other.

This experience was so traumatic that within six months (in March 1984) I gave the Mond lecture at Manchester University on the topic: 'Is Science Manageable?'[2] Coming to the lecture as I did, I was critical of the way the ABRC operated and many of the other members clearly regarded me as very presumptuous – as a mere upstart – in giving the lecture at all. One very distinguished scientist came up to me before the next ABRC meeting and said: 'It's a pity you gave that lecture so soon after joining the ABRC. After another couple of years, you will see that the way we operate is absolutely right.' Having spent almost another four years on the ABRC, I feel that my views, if anything, would be more critical, not less.

To keep readers on the same wavelength as me, I digress for a moment to give one example of the kind of thing that worried me. Whenever a new research possibility came up, the ABRC scientists invariably took the view that, if research in this field was to be carried out anywhere in the world, then it should be carried out also in the United Kingdom. To the annoyance of at least some of them, I coined the phrase 'the British Empire Syndrome' to describe this. I pointed out that, in deciding on the current scale of the British scientific effort, we should not use as our point of reference the British Empire in its heyday. Rather we should ask ourselves what decision would be taken by a rather poor New York State, which is of course about the same size as the United Kingdom and a good deal richer. When I pressed farther and asked how many US centres there were in a particular field of scientific research as compared with the United Kingdom, I was always surprised. Given our relative populations and prosperity, one would expect the United Kingdom to have one-fifth or fewer of the number in the United States. Frequently the answer was a half, or even more. The British Empire Syndrome is manifestly at work.

Of course there is a case for arguing that an independent country should engage in activities on a larger scale than would a state of the American Union. Of course, too, the United Kingdom is a member of some international groupings – for example the European Nuclear Research Centre (CERN) – which enable it to share the burden of research in that very expensive field, and in others. But these are merely qualifications to my main argument.

The implication of all of this is that scientists do not, as an economist naturally would, think automatically of a possible international division of labour. Every economist is taught from his earliest days that no country needs to engage in every field of activity, and especially not in basic research where results are freely published round the world. They may not, as Keith Pavitt would insist, be freely applicable, but they are freely available. So an important, though not necessarily very expensive, part of the United Kingdom's scientific effort should go into monitoring the results of overseas research. This would enable the United Kingdom either to use those results itself or to take up research of its own in fields that international work shows have now become promising. Of course, British scientists do accept the need to maintain a 'defensive' capacity in fields where the main work is being left to other countries but I am constantly surprised at the substantial scale that they claim any such defensive capacity must have. If science is to be managed effectively, those concerned must possess or acquire ways of thinking and acting that every good MBA student acquires, many of them from economics and most of the rest from other social sciences.

But I am getting ahead of myself. For as far ahead as we can see, there will be financial pressure to get maximum value from government spending on science. An international division of labour in science is one way in which we could get better value from the money we spend on it. We could choose to abandon some fields and leave work in them to scientists in other countries, while they chose to leave other fields to us. To say this is to emphasise the need for choice, which is fundamental to all management. Those in charge of any organisation have to choose the direction in which they want the organisation to go and then make subsidiary choices to ensure that it gets there. Science must be managed.

'SOPHISTICATED SIMPLICITY' AND PLANNING

In talking about management, there is one predominant idea which pulls together all that managers do. It is the need for sophisticated simplicity. This may sound naive, but there is a paradox here. Modern organisations inside and outside science are complex, perhaps more complex than most of us think. Yet they can be best managed if their managers can understand clearly, and therefore quite simply, what they are trying to do. We need simplicity in management but, because organisations are complex, we need sophisticated simplicity.

How does this apply to the management of science? At a national level, we can argue whether 'management' is the right word to describe what a body like the ABRC does. At the level of the research institution or laboratory, there clearly is a management job, namely to move it purposefully into the future. Its management must learn to drive a process by which they can obtain sophisticated simplicity.

Those at the head of any organisation need to have a vision of the kind of organisation they would like theirs to be in, say, five or ten years' time. Perhaps this will be the personal vision of the head of the institution, though, if it is, he must somehow ensure at least that the others at the top of the organisation share it. Otherwise he cannot give the organisation effective direction. Better still, the vision for the future shape of the organisation will come from detailed discussion among the senior group in its management – even though this is lengthy and difficult. The sheer process of argument and counter-argument is the best guarantee that the vision really is shared, that it is a vision of 'a future that beckons' for the organisation and those in it. From this vision, the management can go on to work out the organisation's purpose and its mission over the period and from that to establish objectives and strategies for achieving them, so enabling the organisation to write a corporate plan.

The benefit to be derived from corporate planning in scientific research is that it leads to a strong sense of common purpose arrived at through this process of argument and counter-argument which leads to a shared vision of the organisation's future. The resulting vision is simple, and yet inspiring. This explains the recent fashion for producing corporate plans, which are now published by each of the five UK Research Councils and indeed by other bodies concerned with science. Yet, so far, the process of envisioning through planning is incompletely understood. This kind of process is not going on in British science. Though some of its managers will claim they are doing exactly what I am describing, their criterion for seeking and granting research support is too often simply whether this is 'good science', interpreted in university terms.

At the most fundamental level, there is psychological resistance. So the Medical Research Council began its 1986 corporate plan by emphasising that it was ministers and the ABRC who had 'asked' the Council to prepare the corporate plan; by the next sentence this had become an 'instruction'. In a subtle way, the Council distanced itself both from its own plan and from the whole notion of corporate planning. Much of the material in these so-called corporate plans would fit better into annual reports. A plan must establish the future direction for the organisation, not re-travel the past. Indeed one British consultant insists that all verbs in a corporate plan should be in the future tense. That is overstating the case, but it makes an important point.

Some in the Research Councils find corporate planning less valuable than I (and I hope the ESRC as a whole) did. I recall the head of a Research Council complaining at the time taken in constructing a plan since, once it was written, he never looked at it again. 'You amaze me,' I said. 'I look at the ESRC's plan at least once a week.' I looked at it to remind myself what we had committed ourselves to achieve in particular fields and, most important, to judge the appropriateness of proposals for new expenditure, which were steadily being put forward, against the priorities established in the plan. It is precisely the fact that they impose this discipline which generates resistance to corporate plans.

Having experienced at first hand the process of drawing up the ESRC's first plan, I came away even more enthusiastic about corporate planning than 20 years in a business school had led me to expect. I am convinced that there is no better way of ensuring that the broad lines of advance set by an organisation can be achieved. I am also now totally convinced that corporate planning can be effective in non-commercial organisations like those we are concerned with today. Perhaps most of all, corporate plans represent the best way in which an organisation such as a research council or university can demonstrate to those above it in the hierarchy that they are performing well. They can do so without resort to the dottiness of neo-Stalinist performance indicators like those used by the British University Grants Committee.

Another element of complexity in management lies in the fact that all organisations, even smallish ones, inevitably contain large amounts of information – large amounts of what the jargon calls 'variety'. At national level an organisation like the ABRC is confronted by enormous variety – by enormous complexity in the research fields it must cover; in the organisations working in them; in the scientific developments which have to be taken into account in evaluating research proposals; and so on.

No individual could possibly master such complexity. Nor does the ABRC. It is not necessary to ask whether it believes that it could master that complexity, were it to make the attempt, because to my knowledge it has never even considered the issue in those terms. Yet, like the rest of us, the ABRC has clearly devised ways of reducing the variety which faces it, if only by default. Most readers will have heard the remark that the ultimate way of reducing variety is to rely on 'sheer bloody ignorance'. One copes with the complexity of the organisation by ignoring information about it altogether. More realistically, next to that comes the standard device of eliminating much variety by working entirely with overall sums of money, so that enormous amounts of variety are expressed as so many pounds, dollars and so on.

The merit of a good corporate plan is that it reduces variety in a way that clarifies the intentions of the organisation to everyone within it, rather than concealing them. I emphasise that this is not to say that living within the budget and allocating the budget well is unimportant. The point is that a corporate plan composed largely of sums of money – which too many are – cannot give a sense of purpose or direction to anyone. A well-written plan typifies the sophisticated simplicity I advocate. It recognises the complexity of the modern organisation and its environment and yet expresses its intentions in a comprehensible way, making it relatively easy for those in the organisation to plan their work, to do it well and to decide whether to accept proposals which imply developing existing lines of activity or establishing new ones. The plan must reduce variety in a purposeful way, not suppress it. In this spirit I insist that a good plan must be succinct and crystal-clear and must keep references to sums of money to the necessary minimum.

A corporate plan ensures that sufficient time is devoted by all in the organisation to the important developments. It must establish a handful of key tasks for the organisation in the next two or three years, over and above its normal activities. It must go beyond this, too. For example, while the first ESRC plan set a few key tasks to be performed in 1986/7 by the ESRC as a whole, the plan made it possible to establish a further 60 individual tasks. Each was allocated to a specific person in the Council, who was made responsible for its achievement. It will be rare for any organisation to achieve all these key tasks within the allotted period but, provided most of them are, the organisation can ensure that it moves steadily in the intended direction.

The final word on corporate plans is this: it is important to distinguish a plan from a set of motherhood strategies. A plan must spell out what is to happen in the future, not list the achievements of the past, yet, in my experience, this is not what scientists do. Perhaps with the exception of the Agriculture and Food Research Council, what purport to be plans are in reality annual reports. The 1989 Corporate Strategy (not plan) of the Medical Research Council is also a vast improvement on the 1986 document, but there is still too much of the annual report about it.

ECONOMICS AND MANAGEMENT DECISIONS

Establishing the direction in which an organisation should move through corporate planning is obviously important, but once this has been done – and it must be redone every year or two – the task of the organisation is to move forward within the plan. Part of this task will require good decisions to be taken and decision-taking, or problem-solving, will frequently have a significant economic element. If so, a battery of decision techniques can help to make management more effective, most of them deriving from what I would call 'managerial economics'. When I wrote my text book on that subject in the late 1960s I argued that, properly interpreted, managerial economics must bring together a range of decision techniques, many of which were then considered to be part of management accounting or of operational research.[3] This now seems to have happened in both text books and teaching. Managerial economics can, for example, help to allocate the research organisation's resources between different research activities, to control inventories, to assist with costing and pricing – for example for contract research – and to make it easier to decide whether or not to invest in capital assets.

A common feature which unites all these decision techniques is the use of logic – often mathematical logic. An objective is specified, in terms of research grants to be obtained, people to be employed, amounts of research to be carried out and so on. The system for which the decision is to be taken is then similarly specified, usually in the form of a set of interlocking mathematical equations. Logic then comes in, usually to maximise or minimise a mathematical function which specifies the objectives.

I may seem now to have trespassed into the role of mathematics. Nevertheless important economic concepts underlie these decision techniques. To manage science successfully, one must have a sufficient grasp of economics to understand not simply what the decision techniques are doing but, in simpler decisions, to come to the right solution oneself.

I recently spoke to a former student, now a successful investment manager, about what a degree in economics offers in business. He thought it was the ability to 'know in one's bones' what is going right or wrong in an economy or an organisation. I would go further. It is also the ability to know 'in one's bones' what are the right questions to ask when decisions are to be taken. Not that economists are alone in having this ability. In my company, I recently suggested to our UK Managing Director that we might move one of our employees to a job that needed doing. 'No,' he said, 'we pay him too much to let him do that.' His answer came without an instant's reflection because it, too, was in his bones. It was of course also an economist's answer, comparing costs with likely benefit.

What, then, are these basic economic concepts that good managers – including managers of science – should have in their bones? First, and most important, is choice. The resources which scientific and other organisations use – money, people, facilities – are scarce and that is why they have prices. Prices signal to those running organisations that, with given budgets, they can carry out only a limited number of activities. Because only a few of the apparently large number of courses of action available to it can actually be taken, it is necessary for any organisation to choose which activities it will engage in. Drummed into him from his first lecture in his first year economic theory course, the need continuously to choose becomes second nature to an economist. Not so to many scientists, as the British Empire Syndrome demonstrates.

Linked to choice is opportunity cost. Since resources are scarce, if we use them, for example, to seek a cure for AIDS then, within a fixed budget, we have to forgo research in some other field. Or, if the government provides extra funds for the research into AIDS, some other activity which the government would otherwise have financed will have less money. Economists have therefore found that a useful way to measure the cost of undertaking research in a new field is not in terms of money, but in terms of the alternative research opportunity that has to be forgone if the new field is entered. Scientists normally do not, but I am pleased to report that the notion is creeping in. The committee which the ABRC established in 1984 to look at the future of the United Kingdom's contribution to CERN – the European nuclear research facility covered by John Krige in Chapter 6 – looked explicitly at opportunity cost. It tried to discover which extra research in other Research Councils and in universities might have been undertaken had British expenditure on research in nuclear physics been smaller.

SYSTEMS AND META-SYSTEMS

At this point I must pause, because readers may be asking the obvious question. Surely scientists, too, make choices? Yes, of course they do, but they do so in a different context, on their own criteria. To clarify this, we must return to the notion of variety reduction. While managers frequently reduce variety by expressing everything in sums of money, another way they do so is by specialising. Instead of trying to deal with the world as a very complex system indeed, they reduce its complexity by working within sub-systems, which we more usually call 'specialisms'.

This chapter is concerned with three large specialisms – three large sub-systems – science, economics and management. They are shown as circles in Figure 1.1. In a world without specialisation, the three circles would lie on top of each other. In practice, they are largely separate but they are not totally so; there is some overlap. Economists and scientists largely operate within their own sub-systems, which is why I have argued that, for a scientist, the need to choose is not in his bones.

It is important, however, to emphasise that economics is not faultless. Too many economists expatiate about scientific issues only from within their own sub-system. So do managers. All of us, scientists, economists, managers need to concern ourselves with the complete system which embraces all three sub-systems. We may call this the meta-system, the system which is 'over and beyond' the three individual specialisms. It is represented by the total area of Figure 1.1.

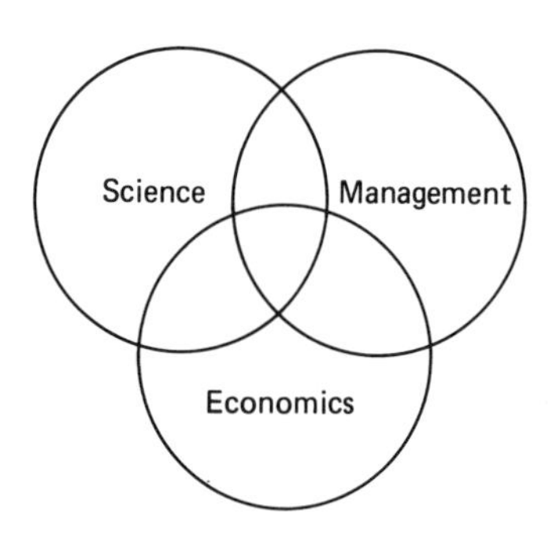

FIGURE 1.1 *The meta-system*

As a preliminary to considering what scientists can offer to the management of the meta-systems of science, let us concentrate on the three areas where two of the sub-systems overlap, using Figure 1.2. The overlap – the intersection – between science and economics gives social worth, with cost and benefit analysis the classic tool for establishing this. To the objection that cost–benefit analysis demeans science by working always in terms of money, the response is that so do scientific budgets. We have to compare like with like.

On the basis of social worth, the intersection between science and management gives us choice. We have seen that choice is in the economist's bones, but so it is in the manager's, as he seeks to get best value from scarce resources. The overlap between management and economics gives us effectiveness. Research projects, once chosen, must be carried out effectively, with those managing them using economic criteria to ensure that resources are used to best effect.

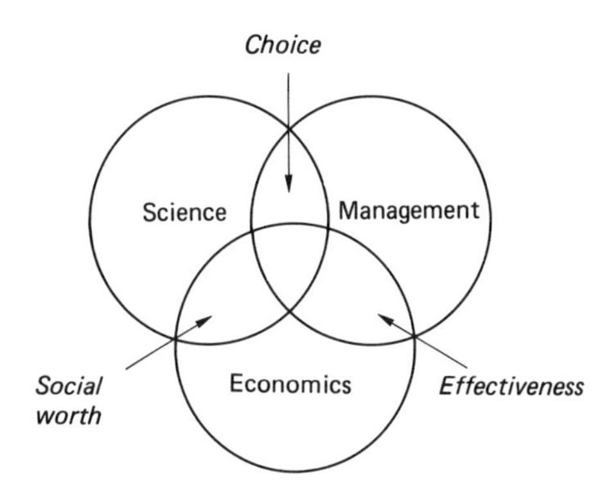

FIGURE 1.2 *Overlapping sub-systems*

The importance of the overlap between science and economics is that again, in his bones, an economist knows the need to look at costs as well as benefits. Costs we all know about, but scientists find it harder to look at benefits. Indeed, in the past, far too many decisions about science – both by scientists and by bureaucrats – have used cost as a proxy for measuring scientific outputs as well as inputs. This procedure drives any economist to distraction and, in John Ziman's 'steady state', scientists also must learn to eschew it.

Because economists are trained to look at benefits and not to use costs as a proxy for them, one of my more annoying habits when Chairman of the ESRC was to pick up a book which was the only tangible result of a major research project and ask, for example: 'Was this worth one million pounds?' I know that there may have been other side benefits from the research but that simply means re-phrasing the question: 'Are all the products of this research project worth one million pounds?' Of course, the answer to such a question is more difficult to give than the question is to ask, but at least we are asking the right question – indeed the only question. Purists will rebel at this insistence on the nice calculation of complex variables. So will those who fear that looking too closely at outputs may very soon lead to the cessation of inputs! Yet this is the only question one can ask in a world of choice, cost and opportunity cost. This may surprise both the purists and those with vested interests, but during my time at the ESRC I found it extremely helpful in coming to decisions about future funding to force ourselves as best we could to answer precisely this kind of question.

For example, in 1985, the ESRC began its move to insist that more PhD students must hand in at least a worthwhile thesis within what I thought was the very generous period of four years. It did so because the records showed that before 1985 about half the PhD students funded by the Council *never* obtained a degree, however long they were given. Our credibility as a Research Council was being undermined and we decided to clamp down. Among the inevitable cries of protest was the suggestion that even those who never acquired a degree at all had learned something from their 'failure'. To that my response was that, if the protesters would care to specify what it was the 'failures', had learned, we could devise a programme specifically designed to enable 'failures' to learn precisely these things. Not only could we do so more quickly and cheaply; having acquired these elusive skills, those in question would emerge from the new programme as successes, not 'failures'. No one ever took me up! But that is how an economist automatically thinks; so should all who manage science.

On costs, accountancy has useful ideas and techniques to offer, but the economist's concept of marginal cost remains crucial to effective problem-solving. In taking any decision involving money an economist will automatically ask: What will this decision add to our costs? He can then balance this against likely extra returns. In other words, he will ask what is the marginal cost which will result from taking the decision and set it against the marginal benefit. Here, surely, you will be saying to yourself, the analysis is being over-simplified. But I promise you, from a lifetime of experience in business, government and universities, that otherwise complex financial calculations are often greatly simplified if we think in terms of marginal costs and benefits.

That same lifetime of experience has led me to conclude that much problem-solving is simply applied common sense but, again, common sense is more rare than one might expect. At Manchester Business School, I often used to say that my title should be 'Professor of

Applied Common Sense', while adding that were this to happen no one would ever think my classes worth attending!

Strictly, the diagrams should be made more complex, by adding a social and organisational sub-system to them. Just as the concepts of economics, and the techniques deriving from them, can make a valuable contribution to the management of science, so can ideas deriving from other social sciences, especially sociology and psychology. Their contribution is more concerned with implementation than with decision-making, but is no less valuable, since their concern is often with the day-to-day organisational activities, like research. As a result of valuable work, especially in the United States and Europe, since the war there is now a large and important body of writing – and a wealth of practical experience – in the successful design and management of businesses and other organisations. This needs to be better applied in the management of science.

Since this is not my own field, and given limited space, I shall not make the diagrams more complex and will say no more about the role of other social sciences. The point is simply that scientists cannot manage science effectively unless they are prepared to understand and apply the lessons which the social sciences offer. If pressed to recommend one book which shows the contribution which the social sciences can make to an understanding of organisations, I would offer *Images of Organisation* by Gareth Morgan.[4]

SCIENTISTS AND THEIR ATTRIBUTES

We have now seen that the social sciences offer scientists many of the techniques and concepts they needed to manage science effectively. In order to do so, however, scientists who manage rather than research must somehow avail themselves of what is on offer. Their difficulty in doing so seems to arise, in part at least, from the fact that the attitudes and attributes of successful natural scientists are not those required of a successful research manager.

When working out what to say here, I asked a colleague with whom I had worked closely in my ESRC days what was the missing attribute which scientists most needed. The reply came without a moment's thought: humility. There is unfortunately a fine line between claiming to possess high specialist skills and sheer arrogance. There is also a fine line between an insistence that one's own subject is that most deserving of financial support and a determination to insist that the support must be made available. As a senior American scientific administrator said to me: 'The nuclear physicists would spend the whole of the GNP on research in nuclear physics if we would let them. So would the astronomers but, being nicer guys, they would be prepared to go 50/50 with the physicists.' And that remark is only partly a joke. In the United Kingdom, for example, nuclear physics spends at least twice as much on research and on the training of research students as do all the social sciences together.

The rest of us do, of course, have doubts about the beneficence of nuclear physics, for all kinds of reasons. Perhaps it is medical research which now obtains the biggest advantage from public attitudes. Medical research is clearly important, and it is not easy for people outside medicine to turn down proposals for extra expenditure which they should turn down, if told that to do so will endanger human life – even their own. Never believe a medical man who tells you that his proposals for research funding do not obtain a fair hearing from others. The criteria enunciated in the previous two sections are just as relevant to decisions on spending on medical research as in any other field. Perhaps they are more so, given the huge charitable funds available in the medical field.

Beyond these issues of how far to support particular areas of science or medicine, there are questions about support for science as a whole. 'Lack of humility' on the part of scientists has its effects here, too. While most scientists still insist that more should be spent on science, both the public and some politicians have increasing doubts. There is no need to rehearse the issues which have led to the development of the green movement and to its calls for slower, perhaps even negative, economic growth. But if the sustainability of economic growth is to be called into question, then so is that of continuing increases in expenditure on science.

Even if there is still room for argument over whether the development of science should actually be held back, there are strong grounds for arguing that science and technology have been allowed to develop beyond our ability to manage them for the benefit of society as a whole, or of individual organisations within it. We need to shift the balance between the effort put into new developments in science and technology and that going into understanding their social implications and discovering how to manage them better.

Of course there are scientists who do worry about expenditure on science, though their voices are rarely heard when decisions on the amount of public money to go into science are made. Dr Peter Higgs, who was responsible for setting off the expensive search for the elusive 'Higgs boson' has been reported as saying that he is bemused by the vast resources thrown into the hunt for the particle he predicted back in the 1960s.[5] 'Of course I am flattered by it all,' he told an American magazine. 'Of course I like it when they give seminars on the search for the Higgs. And of course I think it's important to look for. But – you want to know the truth? When I consider the huge sums going for this, the lifetimes spent on research, I cannot help but think "Good heavens, what have I done?"'

The problems which scientists and technologists face in applying their specialist talents to the management of laboratories or other organisations is that these are meta-systems and that the economic, and indeed the social and organisational, sub-systems which they contain are very different from the systems which scientists study. The problem does not particularly lie in the problem of predicting how meta-systems and their environments will change, since the behaviour of scientific systems can be difficult to predict as well. The cardinal difference between the organisational meta-system and its scientific sub-system is the preponderant role played in meta-systems by people, whether individuals or groups. An understanding of how human beings will act and react in different circumstances – let alone how they can be managed in these circumstances – calls for an understanding of the social and organisational sub-system, which in turn calls for expertise in fields such as sociology, social psychology and anthropology, which few natural scientists possess, or feel the need to possess. To put that expertise into effect also calls for a good deal of practical managerial experience.

I do not suggest that the solution is for scientists to become social scientists as well. In part, this is because seeking to master both a field of science and many of the social sciences as well would require an investment of time and an ability to think in a diversity of ways that few but supermen could possess. (The vast increase in the sum of relevant human knowledge over the past half-century has made the polymath an endangered species.) In part, however, it is because it is not necessary to become a polymath. The problem of having a big enough spread of expertise to manage the meta-system can be overcome if an activity is managed by a team whose members between them have sufficient scientific, economic, social scientific and managerial knowledge and experience.

What happens then is shown in Figure 1.3, which points to the intersection between all the three sub-systems of science, economics and management. Where these overlap, one has objective research plans, and the bigger the intersection between the three sub-systems the more valuable the research plans will be. To be certain of a large overlap we must ensure that the whole range of relevant skills is deployed.

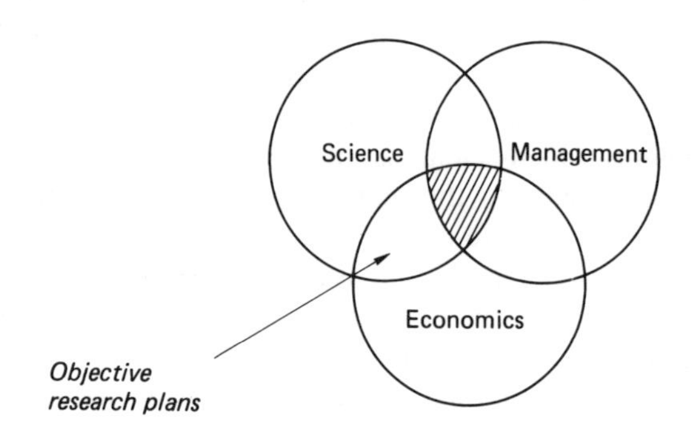

FIGURE 1.3 *Objective research plans*

But it must be deployed. What worried me most about the ABRC was the amateurism with which it too often operated, when concerned with issues of management or organisation. There was not enough professionalism there. The ABRC is run on 'club' lines – almost as an offshoot of the Royal Society – so that entry is by scientific eminence; nevertheless the professionalism that it lacked should have been found. To behave unprofessionally when a professional approach is available is unacceptable. It implies that one's innate talents can tackle any management problem. It demonstrates what Norman Strauss calls 'untutored arrogance'.[6] I must, however, make it clear that I have enormous respect for the way in which, as Chairman of the ABRC, Sir David Phillips has pushed his colleagues towards more effective policies. But even he could not introduce into the way it operates the professionalism which the ABRC needs.

The answer to the question whether scientists can and do possess the attributes and skills needed to manage science is therefore threefold. First, most scientists have the ability to learn the skills both theoretically and practically, though it is questionable whether the time required would be well spent, not least because the meta-systems which managers have to manage are different from the sub-systems which scientists investigate. Second, scientists can avoid the necessity to become expert over this wide range of knowledge and expertise by working, in teams, with those who possess the skills which they do not. Indeed, given that most of us are neither supermen nor polymaths, the only way that we can manage the complexity of modern organisations is to recognise that *team* leadership and *team* management is essential. Third, a limited number of scientists and technologists seem by some automatic process of osmosis to have acquired enough understanding of management practices and processes to manage the meta-systems of science well. But they are very rare birds. The rest of us would be wiser to assume that we do not fall into that category.

THE ROLES OF GOVERNMENT AND BUSINESS

Finally, what role should government and business play in the management of science? It should by now be obvious why they have a role. It is partly because a scientific training does not normally provide even an adequate introduction to the aspects of social science which are relevant to management. It is also because a career in science normally allows only exceptionally able – or very fortunate – scientists to acquire a satisfactory understanding of the principles and processes of management on the job. It is partly, however, because society must inevitably find it unacceptable for scientists alone to bid for, and share out, research money, with no umpire in sight.

The obvious source of such umpires is business or government. The problem then is that the umpires' own backgrounds are themselves inadequate to the task. This is because one can be a successful umpire only if one has a good understanding *both* of science *and* of the systems to be managed. Unfortunately, most businessmen and most civil servants lack important attributes that a complete umpire would have.

The problem which businessmen face arises precisely from their very success in reducing the complexity of the business and management problems they face, in order to tackle them successfully. Especially at the Whitehall level, they soon see that the problems of business actually look rather simple when compared with those of government. Or, if they do not, that can only make matters worse. Businessmen can make a contribution in a wider group because of the way in which they simplify the issues but, left to themselves, businessmen would over-simplify the complexity of managing science.

Civil servants are different, and a major reason why they are is that they have grasped the complexity of the issues with which they are concerned. Solutions to problems of policy or management have to be fair but that fairness must be demonstrable to ministers, to parliament and to the public. As a result, while a businessman might suggest solutions to problems which he would see as managerially efficient, civil servants would rule them out because they would contravene basic principles of British constitutional practice. Civil servants *will* find constitutionally acceptable solutions, but they, in turn, may well be managerially inefficient.

The kind of problem which arises can be seen in the activities of the ABRC. Inevitably, bodies which debate and advise on national policies for science have to be established by government, even though the bodies themselves may seek to influence private businesses as well as the public sector. The result is that the process by which national policies for science are determined is run on civil service lines. Best managerial practice is rarely used. This is the case, for example, with the size of the ABRC. The civil service case for size is, above all, that the easiest way to keep all interested parties informed is to let them sit in on meetings, even though the size of these meetings frequently means their members sitting for several hours through interminable discussions while 20 people say their pieces. The cost of an ABRC meeting is about 100 valuable man-hours. I have little complaint about the spread of backgrounds of ABRC members or of their specialist abilities. My complaint is that there are too many of them at each meeting for any individual to feel that he is spending his time fruitfully and that equally few of them understand management.

Most of all, the heads of all five research councils are ex-officio members of the ABRC and therefore have – and should recognise that they have – conflicts of interest. Should they support a decision because it is in the national interest? Or should they oppose it because it would divert resources from their own Councils? Too often, the answer is the latter.

It is also clear from the way that bodies like the ABRC operate that civil servants rarely ask what a particular meeting is for, and therefore rarely consider how it should be organised. Taking the ABRC as an example: is it a debating chamber, an information exchange, or a body advising the Secretary of State on how big a science vote he should seek from the treasury; and on how it should be divided between the research councils and other research bodies? I would myself want the Board to be a small group whose role was to adjudicate between conflicting claims for government money and to find less time-consuming methods for its role of exchanging information. I must, however, point out at once that the 'independent' members of the Board already hold closed meetings, when important stages in the decision-making process are reached and it seems better for interested parties to be excluded. But perhaps they should always do this.

This is to get into too much detail. These questions are raised simply to make quite clearly the point that, however academic and business thinking on managerial issues changes, the civil service machine goes on operating in time-honoured ways. It does so because civil servants spend too little of their time thinking and learning about organisational processes and because those who do think about them are rarely promoted to the interesting or glamorous positions in the service.

However we can now raise this discussion to a higher level. If national science policy is to grasp the opportunities for change, in a world where developments in high technology are crucially important to economic development, a body like the ABRC needs to work out how best to spark off and manage innovation. It must become a learning organisation – learning how to innovate both in ensuring that the research it funds is forward-looking and innovative and in ensuring also that the management of its own organisation is the same.

CONCLUSIONS

The title of this chaper is a question: Can scientists manage science? I believe I have shown that they can do so effectively only if either they themselves acquire a practical understanding of the social sciences, not least, of economics and of management; or if they work in inter-disciplinary teams which include and value those who do have such knowledge.

Unfortunately, as has been shown in the last part of this chapter the civil service remains the joker in the pack. Civil servants remain at least as ignorant as scientists of the problems and possibilities of modern management. Yet, so long as a high proportion of the money going into science comes from public funds, civil servants will continue to have the biggest influence, if not on the way in which those funds are allocated at least on the organisational structures within which that happens. Given the nature of the civil service, for the forseeable future perhaps the best that the management of science can hope for is still the dawning of a better yesterday.

May I add a postscript – though not an afterthought? I passionately believe that there must be a step change in the effectiveness of the management of British science, and this can occur only if scientists act on what I have said in this chapter. Equally passionately, I doubt whether any of them will. I therefore simply add this. One of the ways in which society should judge the fitness of science to be given society's resources in future must be whether any notice is taken of the kind of idea I have discussed. In that sense, science will judge itself.

NOTES

1. John Ziman, *Science in a 'Steady State'*, available from the Science Policy Support Group, 22 Henrietta Street, London EC2 8NA
2. Sir Douglas Hague, *Is Science Manageable?* Mond Lecture, University of Manchester, 12 March 1984. Copies available from Sir Douglas Hague, Templeton College, Oxford.
3. Sir Douglas Hague, *Managerial Economics* (London: Longmans, 1969).
4. Gareth Morgan, *Images of Organisation* (Beverly Hills, London/New Delhi: Sage Publications Ltd).
5. *Sunday Times Magazine* (London) 2 July 1989.
6. Norman Strauss is co-director, with Sir Douglas Hague, of the Strategic Leadership programme at Templeton College, Oxford

2 What Do We Know about the Usefulness of Science? The Case for Diversity

KEITH PAVITT

In this chapter, I discuss the contribution (past, present and potentially in the future) of economic analysis and other social sciences to the management of science. I shall restrict myself to strategic management – the procedures, processes and criteria whereby resources are allocated to research activities, progress is monitored and results are disseminated and assessed – rather than to what goes on inside the laboratory. I shall also concentrate on science rather than technology, although we shall see that the nature of the contribution of the former to the latter is of central importance.

I shall argue in the opening section that increased analytical interest in the better management of science is justified by its growing importance, rather than by the disputable assertion that it is not growing any more. In the second section I identify two major contributions by (political) economists to better understanding: the growth of science as a factor of production, and the economic case for the public subsidy of basic research. I argue that unwarranted emphasis by contemporary economists on the 'public good' and information-like properties of science (and sometimes even of technology) has led to the neglect of two centrally important problems of contemporary science policy; the contribution of science to technology (discussed in the third section), and the supposed 'internationalisation' of science and technology (fourth section). Next I then argue that economists can make a useful contribution to better understanding of the properties of basic research-performing institutions. Finally on the basis of empirical research on these problems, I conclude that (amongst other things) there is no case for 'greater selectivity and concentration' in British basic research, and, as a postscript, I argue that the contribution of analysis to science policy is no different from the contribution of science to technology.

STEADY STATE SCIENCE?

It is commonly argued that we need better knowledge related to the management of science, given the agonising choices that must be made, now that science has entered a 'steady state'. The issue of a steady state of zero growth in science was first raised in the public consciousness by Derek de Solla Price in his classic essay 'Little Science, Big Science'. published in 1963. Some analysts say that it has now happened, or is about to happen (see, for example, Ziman, 1987). However the available data suggest otherwise.

Figures 2.1 and 2.2 show that, at least since the mid-1960s, civilian R&D has grown in the OECD (Organisation for Economic Co-operation and Development) area, both in real terms and as a percentage of GDP. Growth has been particularly rapid in Japan, and slow in the United Kingdom. After a deceleration in the 1970s, the rate of growth has in fact increased in the 1980s. Furthermore data published by the National Science Board (1987, p.52) show that the total employment of scientists and engineers in the United States increased annually by 6 per cent between 1976 and 1986, and is expected to increase by a further 36 per cent by the year 2000. In Japan, the numbers of science and engineering graduates continue to increase and, as in the United States, a growing proportion are finding employment outside manufacturing in professional services, finance and insurance (Kodama, 1989). These are not symptoms of a stationary state, but rather a vigorous growth in demand for professional and research skills in science and technology.

For the science base, strictly defined, the picture is more complicated. According to OECD estimates, R&D in higher education grew in real terms in all OECD countries until the mid-1970s, when it stabilised and even declined slightly in West Germany, Italy and the United Kingdom (OECD, 1980, p.44). According to J. Irvine *et al.*(1986), real growth in academic science between 1975 and 1982

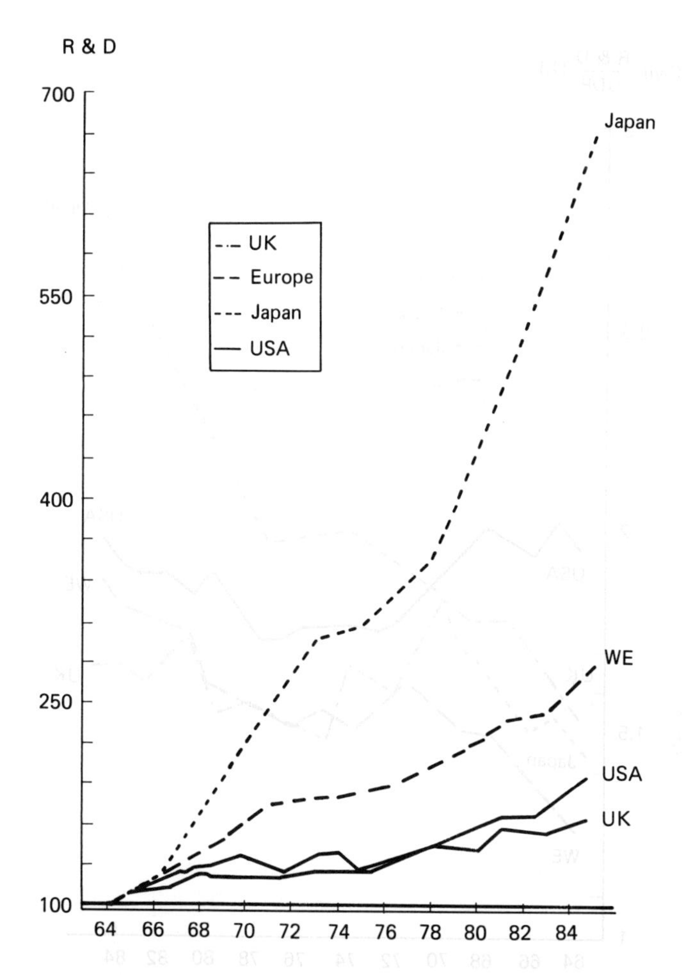

FIGURE 2.1 *Trends in real expenditure on civilian R&D (1980 prices and purchasing power parities 1964 = 100)*

Note: Europe (WE) comprises Belgium, France, FDR, Italy, Netherlands, Sweden, Switzerland, UK.

Source: OECD Science and Technology Indicators Unit.

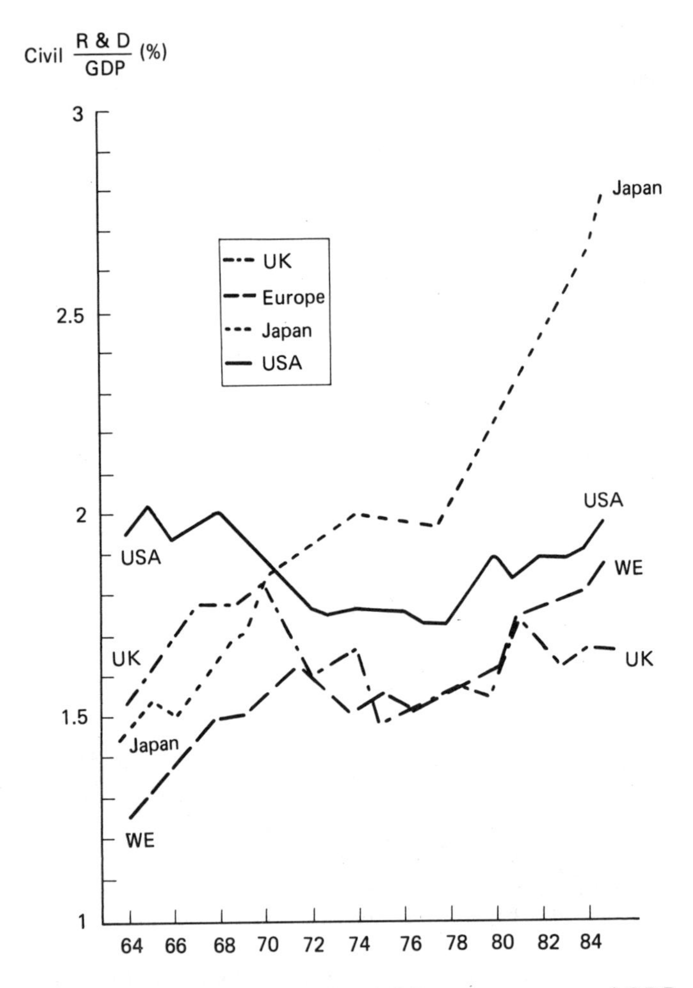

FIGURE 2.2 *Trends in civilian R&D as a percentage of GDP*
Source: As Figure 2.1.

was slow in all major OECD countries except Japan and France. Together with P. Isard and B. Martin, Irvine will shortly be telling us what has happened since then (1989). Amongst other things they will show a buoyant trend, in most major OECD countries, in funds specifically earmarked for basic research, compared to the notional research component of general university funds.

In the meantime, one thing is clear from all the comparisons: the low levels and rates of growth of UK R&D expenditures. Together with the comparisons of basic research by Irvine and Martin, Figures 2.1 and 2.2 show that Britain is closer to a steady state than other OECD countries. Even if other countries' R&D expenditures do eventually stop growing, they will be at higher proportions of national resources. In other words, the United Kingdom is heading for a 'low-level steady state', whereas other countries will find themselves at a 'high-level steady state'.

Thus the evidence suggests that, for most countries, the management of science takes place in a regime of growth in demand for research skills and knowledge, no doubt reflecting the importance of technological change in economic efficiency and welfare (see, for example, Fagerberg, 1987, 1988). The case for improving the knowledge-base for the management of science must therefore be the old-fashioned one of 'timeliness and promise' It is timely to improve understanding of an activity that consumes considerable resources (approximately 0.3–0.5 per cent of GDP) and that has a major influence on society's capacity to respond to economic and social demands. And there is considerable promise of such improvement, given our qualitative understanding of the nature and determinants of science and technology (Mowery and Rosenberg, 1989) and recent advances in quantitative, bibliometric techniques. In this context, I shall suggest a number of subjects where better understanding could help improve the management of science. But before this, I shall briefly review some of the past contributions by economists.

SOME CONTRIBUTIONS BY ECONOMISTS

Economic thinking has made two major contributions to the policy debate about the management of science. It has shown, first, that the growth of basic science must be understood and justified mainly by its contribution to economic and social progress and, second, that basic science should be supported mainly through public subsidy.

Science as an Economic Activity

The importance of science as an economic activity was in fact recognised very early on. In Chapter One of *The Wealth of Nations,* Adam Smith pointed out that technical advances were made not only at the point of production, but also by suppliers of capital goods, and by 'philosophers and men of speculation', which is what scientists were then called. Perhaps less well known in this country are the predictions of Alexis de Tocqueville in *Democracy in America.* He observed the down-to-earth and problem-solving nature of US society early in the nineteenth century, and predicted the rapid expansion of science, for three reasons: first, as a form of conspicuous intellectual consumption, funded from the great accumulations of private wealth that de Tocqueville rightly predicted the US system would produce; second, as a foundation for the education of the large number of applied scientists that de Tocqueville predicted (again rightly) modernising society would require; and third, as a source of fundamental knowledge needed to facilitate and guide the solving of practical problems.

Thus the rapid growth of modern science must be seen as part of the more general process of the specialisation and professionalisation of productive activities in modernising societies. To this we must add Marx's important insights into the major influence of economic and social demands on the rate and direction of scientific advance, through the problems that they pose, the empirical data and techniques of measurement that they generate, and the financial resources that they make available (see Rosenberg, 1976). For all these reasons, economists are right to argue that large expenditures on science cannot be understood and justified solely on cultural and aesthetic grounds; they inevitably have important economic and social dimensions. However we shall see in the following section that the links today between science and technological practice are far from simple or straightforward.

The Public Subsidy of Science

In the meantime, another major contribution of economics to the management of science has been the analytical justification for regular and large-scale government funding of basic research. As is often the case, principle followed practice rather than leading to it, since governments in some countries (and most notably Germany) had already been funding basic research for a very long time. After the Second World War, the United States followed suit, and in the early 1950s, established the National Science Foundation. In 1959, R. Nelson published his pioneering paper entitled 'The Simple Economics of Basic Scientific Research', in which he argued that, left to itself, a competitive market will invest less than the optimum in basic research. This is because a profit-seeking firm cannot be sure of capturing all the benefit of the basic science that it sponsors, given major uncertainties about the benefits for the sponsoring firm, and the difficulties it faces in extracting compensation from subsequent imitators. At the same time, a policy of secrecy aimed at stopping such imitation would be sub-optimal, since it would restrict applications with small marginal cost. If, in addition, profit-seeking firms are risk-averse, or have short-term horizons in their decisions to allocate resources, private expenditures on basic research will be even more sub-optimal.

Nelson's insights have been developed and modified over the past 30 years, notably by Arrow (1962) and Averch (1985) in the United States, and by Kay and Llewellyn Smith (1985), Dasgupta (1987), Stoneman (1987) and – most recently – Stoneman and Vickers (1988) in the United Kingdom. Risk-aversion, low or zero marginal cost of application and difficulties in appropriating benefits have become standard explanations for the public subsidy of science. At the same time there has been a subtle shift in emphasis. Nelson's original paper was grounded in research on the development of the transistor (1962), and his paper is spliced with examples of the development and application of science. Over time progressively fewer references have been made to empirical evidence, and more to the standard theorem of welfare economics. Whilst it might be advantageous in the economics classroom to stress the 'public good' characteristics of science, and to minimise or ignore the distinctions and interactions between science and technology, this has effectively excluded economists from two of the major debates of contemporary science policy: the nature and extent of the contributions of science to technology, and the impact of national science on national technology (Mowery, 1983).

Technology as Science?

It is comfortable as well as convenient to treat science and technology as the same thing, given the similarities in their inputs (scientists, engineers, laboratories) and their outputs (knowledge) and given the well-known examples of outstanding science performed in corporate laboratories. However this neglects the very different nature and purpose of the central activities of university and business laboratories. In universities, basic research seeks generalisations based on a restricted number of variables, and results in publications and reproducible experiments. In business, a combination of research and (more important) development, testing, production engineering and operating experience accumulates knowledge on all the critical operating variables of an artefact, and result in knowledge that is not only specific, but partly tacit (uncodifiable) and therefore difficult and costly to reproduce.

Given these differences, basic research is more likely to meet the conditions for private under-investment, as defined by Nelson and others, which explains the higher proportion of public funding in basic research than in development in all OECD countries. Economists conscious of the distinction between science and technology have made a major contribution to the policy debate by stressing the complementary nature of private and public investments in science and technology, with the former concentrating on the short-term and specific, and the latter on the long-term and the general. They have also warned of the dangers and inefficiencies of heavy public funding of commercial development activities (Eads and Nelson, 1971; Jewkes, 1972). However insufficient attention has been directed by economists to the interface between science and technology.

Science as a 'Free Good'?

One other reason for this lack of attention has been a common confusion between the reasonable assumption that the results of science are a 'public good' (i.e. codified, published, easily reproduced and therefore deserving of public subsidy) and the unreasonable assumption that they are a 'free good' (i.e. costless to apply as technology, once read). In a paper entitled 'Why do Firms do Basic Research (with Their Own Money)?, N. Rosenberg (1990) argues that basic research financed and performed in (mainly large) firms often grows out of practical problem-solving, and that the two are highly interactive. He also argues that in-house basic research is essential in order to monitor and evaluate research being conducted elsewhere:

> This point is important...in identifying a serious limitation in the way economists reason about scientific knowledge and research in general...such knowledge is regarded by economists as being 'on the shelf' and costlessly available to all comers once it has been produced. But this model is seriously flawed because it frequently requires a substantial research capability to understand, interpret and to appraise knowledge that has been placed on the shelf – whether basic or applied. The cost of maintaining this capability is high, because it is likely to require a cadre of in-house scientists who can do these things. And, in order to maintain such a cadre, the firm must be willing to let them perform basic research. The most effective way to remain plugged in to the scientific network is to be a participant in the research process.

This has implications for the way we view the impact of science on technology, and for the reasons for public subsidy. We shall take them up in the following three sections.

THE IMPACT OF SCIENCE ON TECHNOLOGY

The impact of science on technology is bound to be of central concern to science policy-makers. I summarise below what we already know from earlier studies, and identify subjects for future research.

Calculating the Economic Return from Basic Science

Some resource-starved British scientists would no doubt strongly welcome a study demonstrating a high economic return to basic research, and leading to unaccustomed munificence from an economically hard-nosed Treasury. In fact, one such study has just been completed in the United States by a distinguished economic expert on R&D, E. Mansfield (1989). It is one of the most ingenious and persuasive of its kind but, as pointed out by P. David and his colleagues (1988), calculations of this kind do not satisfactorily reflect the nature of the impact of science and technology:

> The outputs of basic research rarely possess intrinsic economic value. Instead, they are critically important inputs to other investment processes that yield further research findings, and sometimes yield innovations ... Policies that focus exclusively on the support of basic research with an eye to its economic payoffs will be ineffective unless they are also concerned with these complementary factors.
>
> The alternative conceptualization ... that we have developed focuses on basic research as a process of learning about the physical world that can better inform the processes of applied research and development. Rather than yielding outputs that are marketed commercially, basic research interacts with applied research in a complex and iterative manner to increase the productivity of both basic and applied research. The development of links between the basic and applied research enterprises are critical to the productivity and economic payoffs of both activities. (pp. 68–9)

The Complexity of Science's Impact on Technology

We know from the results of past research that the links between basic science and technology are in fact complex along at least four dimensions:

1. The intensity of *direct transfers of knowledge* from basic science to application varies widely amongst sectors of economic activity, and amongst scientific fields. The most systematic analyses have been made in the United States, on the basis of patent citations to journals (Carpenter, 1983; Narin and Noma, 1985) and of a survey of industrial R&D directors (Nelson and Levin, 1986). They both confirm strong links in chemicals and drugs firms to basic research in biology, whilst the links of electronics firms are also intense, but to more applied research activities, in physics. In mechanical and transport technologies, on the other hand, the links to science are weak.
2. The *impact of basic research on technology* also varies widely from the generation of epoch-making new technologies (e.g. electricity, synthetic materials, semi-conductors; see Freeman *et*

al., 1982), through the routine chemical analysis for quality control in continuous flow industries (Rosenberg, 1985), to insights and methods for dealing with applied problems. In all cases, operationally viable technology requires combinations with knowledge from other sources, including design and production engineering.

3. Basic science has an impact on technology, not just through direct knowledge transfers, but also through *access to skills, methods and instruments* (Pavitt, 1987).
4. Knowledge transfers are mainly *person-embodied*, involving personal *contacts, movements and participation* in national and international *networks* (Gibbons and Johnston, 1974).

Is Basic Research a Growing Source of Technology?

Some analysts (for example, Martin and Irvine, 1989) claim that we are now witnessing a significant increase in the direct use in technology of the results of basic research. Others claim that such 'strategic' areas of science should receive priority support from government. In my view, the evidence is ambiguous and incomplete.

Narin and Frame (1989) have produced the most persuasive quantitative evidence so far. Figure 2.3 shows sharply upward trends in the frequency with which US patents, originating in a number of countries, contain citations to publications other than patents. On this basis they claim that the technology reflected in US patents is much more 'science-dependent' than ten years ago. They further show that the time-lags in the citations from patents to other publications are diminishing rapidly, and that science-intensive patents are relatively highly cited.

Whilst it is suggestive, this evidence has its limitations. It is not yet clear to what extent the 'other publications', cited in patents, reproduce basic or applied research, from universities or from corporate laboratories. In addition, a high proportion of technology is not patented, because it is kept secret (e.g. process technology) because it is tacit and non-codifiable know-how, or because – as in the increasingly important case of software technology – it is very difficult to protect through patenting. This non-patented technology is likely to be less dependent on science, and more on cumulative design and engineering skills. Together with a number of colleagues,

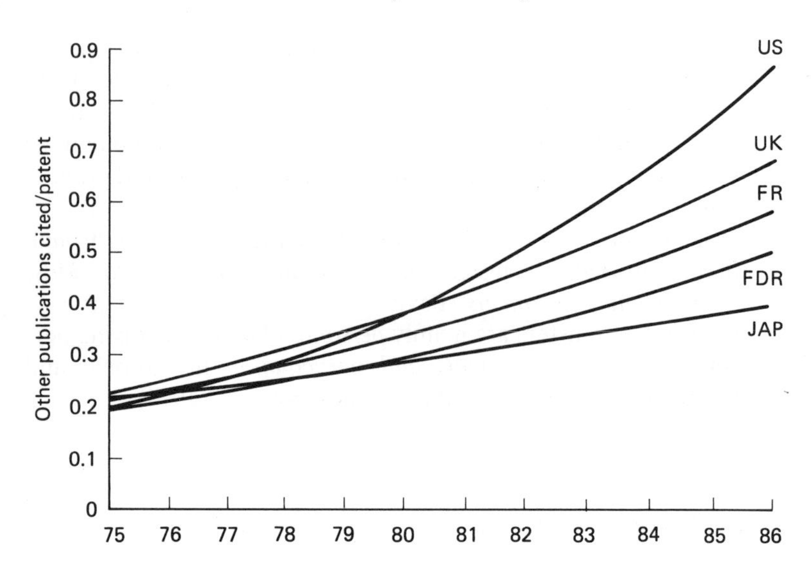

FIGURE 2.3 *Patent science intensity for five countries (smoothed)*
Source: Narin and Frame (1989).

I have argued elsewhere that it is increasing as a proportion of total technological activity (Soete *et al.*, 1989).

In addition, it is worth noting that, in the United States, the recent report from MIT *Made in America* (1989) has claimed that it is precisely because of deficiencies in these engineering skills that US firms are not capturing the full economic benefits from exploiting scientific advances. They further claim that engineering education in the United States has become too science-based.

More generally the evidence from US R&D statistics is ambiguous. Whilst there are signs of increasing corporate commitment to basic research in the 1980s, this follows an extended period of decline, and it has only just regained its share of the early 1960s (National Science Board, 1987, Appendix Tables 4–10 and 5–1). No systematic data are available on trends in corporate support in different fields of science. This would enable a confirmation or otherwise of an explanation in terms of cycles of rich applicability of specific fields of academic science: for example, solid-state physics in the 1960s; biotechnology in the 1980s.

Even if certain fields of basic research are increasingly important in their direct knowledge inputs into technology, it is misleading to assume that only they contribute to technology, and other fields do not. There are at least two other influences of science and technology that are equally – if not more – important: research training and skills; and unplanned applications.

The Broad Demands for Research Skills

One important function of academic research is the provision of trained research personnel, who go on to work in applied activities and take with them not just the knowledge resulting from their research, but also skills, methods and a web of professional contacts that will help them tackle the technological problems that they later face.

In one of their less well-known studies, Irvine and Martin (1980) have shown that masters and doctoral graduates from British radio-astronomy benefited in subsequent non-academic careers from the research skills – rather than the research knowledge – that they obtained during their post-graduate training. A more comprehensive survey undertaken at Yale University suggests that this is the rule rather than the exception.

The relevant results are summarised in Table 2.1. They show the responses of 650 US industrial research executives, spread across 130 industries, who were asked to rank the relevance to their technology of a number of fields of pure and applied science. Table 2.1 lists the number of industries in which each scientific field was given high ranking according to two criteria: first, the relevance of the *research skills* in the science to the technology; second, the relevance of the *academic research* in the science to the technology. As the authors point out:

> Industrial scientists and engineers almost always need training in the basic scientific principles and research techniques of their field, and providing this training is a central function of universities. Current academic research in a field, however, may or may not be relevant to technical advance in industry, even if academic training is important. (Nelson and Levin, 1986)

Table 2.1 shows that, in most scientific fields, whether pure or applied, academic training and skills are relevant over a far larger number of industrial technologies than is academic research. The exceptions are the pure and applied biological sciences, where we

TABLE 2.1 *The relevance of scientific fields to technology (USA)*

Field of science	*Number of industries (out of 130) ranking scientific field at 5 or above (out of 7) in relevance to its technology of:* Science (skills)	Academic research (knowledge)
Biology	14	12
Chemistry	74	19
Geology	4	0
Mathematics	30	5
Physics	44	4
Agricultural science	16	17
Applied maths & operational research	32	16
Computer science	79	34
Materials science	99	29
Medical science	8	7
Metallurgy	60	21

Source: Nelson and Levin (1986); Nelson (1987).

know from other studies that academic research is at present very close to technology (Narin and Noma, 1985). These results show clearly that most scientific fields are much more strategically important to technology than data on direct transfers of knowledge would lead us to believe.

Unplanned Applications

Another major influence of science on technology is through unplanned applications, where useful knowledge emerges from research undertaken purely out of curiosity, without any strategic mission or expectation of application. Two US studies – one undertaken in the late 1960s and the other some 20 years later – show the importance of such research for achieving relatively short-term technological objectives (Illinois Institute of Technology, 1969; Mansfield, 1989). In both cases, important innovations would have been substantially delayed without contributions from unprogrammed research performed in the ten years preceding the commercial launch of the innovations. Furthermore, in both studies, unprogrammed research contributed about 10 per cent of the important knowledge inputs.

One implication of these findings is that programmed R&D should be built on a wider spread of non-programmed research. Analysts such as Nelson (1959) and Kay and Llewellyn Smith (1985) have gone further and used various examples to suggest that more useful knowledge is produced in the long term by allowing basic scientists to pursue their own interests, than by fixing practical objectives for their work. It is a view that needs to be considered seriously by analysts in the future (see, for example, Council for Science and Society, 1989).

IS SCIENCE (AT LAST) BEING INTERNATIONALISED?

The analytical apparatus described above has in general assumed a closed economy. This is paradoxical, given that the main stimulus for public policies for science and technology has in fact probably not come from any notions of (national) market failure, but from what is perceived as best practice in a world system of international competion where technological leads and lags are of central importance. It is also perhaps fortunate that the subject has not been pursued too often within the mainstream analytical framework: if we assume that basic research is a 'free good', an open international world would, in principle, permit any one country to live off the rest of the world's basic research (Kay and Llewellyn Smith, 1985).

But the real world is more complicated. As Rosenberg (1990) has pointed out, the ability to assimilate the results of other people's basic research depends in part on performing basic research oneself. An active national competence in basic research is therefore a necessary condition for benefiting from research undertaken elsewhere in the world; indeed it can be viewed as a national scientific intelligence system. And since most transfers of knowledge and skills between science and technology are embodied in individuals, the constraints of distance and language have meant that nation-based transfers between science and technology have been the exception rather than the rule.

Now, it is argued, conditions are changing. The barriers of distance and language are lower than they used to be. And firms are increasingly internationalising their R&D activities, which enable them to assimilate the results of academic science in a number of countries. Does this mean that linkages between science and technology will become internationalised? Does it mean that an increasing proportion of the benefits of national governments' investments in basic research will 'leak away' through multinational firms to other countries?

This is a subject that deserves further research. Suffice it to suggest at this stage that the degree of 'leakage' depends, as a first approximation, on the proportion of a country's corporate technological activity that is controlled by foreign firms, which reflects their capacity to monitor and absorb local basic research. Similarly, the importance of the foreign technological activities of nationally-owned firms will reflect a country's capacity to benefit from the results of basic research undertaken in other counties.

Table 2.2 is a first attempt to measure and compare these variables across countries. The first column compares the proportion of each country's US patenting originating from foreign-controlled firms. It shows that, in most countries, large foreign firms still play a relatively small role in national technological activities; only in Belgium, Canada and the United Kingdom do they account for more than 20 per cent of the total. The second column compares the US patenting of nationally-controlled firms from outside their home country, as a proportion of total national patenting in the United States. It shows,

TABLE 2.2 *Foreign-controlled domestic technology compared to nationally-controlled foreign technology (based on US patenting, 1981–6)*

Home country	*US patenting from inside country by foreign firms (as % of country's total US patenting)*	*US patenting by national firms from outside home country (as % of country's total US patenting)*
Belgium	45.7	16.5
France	11.8	3.8
FDR	11.5	8.5
Italy	11.2	3.0
Netherlands	9.5	73.4
Sweden	5.4	16.7
Switzerland	12.5	27.8
UK	22.3	24.5
W. Europe	7.4	9.3
Canada	28.1	12.5
Japan	1.2	0.5
USA	4.2	4.4

Source: Patel and Pavitt (1989b)

as might be expected, that large firms based in small countries undertake a higher proportion of their technological activities outside their home countries. For the Netherlands, they amount to more than 70 per cent of the national total, and more than 20 per cent for Switzerland and the United Kingdom. Given its size, the proportion is relatively high for the United Kingdom.

Taken together, the two measures show that most national technological systems are relatively self-contained. Both measures of internationalisation are less than a quarter of total technological activities in eight out of the 11 countries. In Belgium and Canada, foreign-controlled domestic technological activities are much greater than domestically-controlled foreign technological activities, whereas for the Netherlands and Sweden the opposite is the case. When Western Europe is considered as a whole, the degree of internationalisation is much less than for most of the European countries taken individually, but still greater than that of either Japan or the United States.

These results show that complete internationalisation of links between science and technology is not at all likely in the immediate future; in most countries, national science will still be feeding into largely nationally-controlled technology, and close links with foreign science will in most cases be small compared to national links. The degree of internationalisation of the United Kingdom is greater than for the other European countries of equivalent size, but is still less than a quarter. Contrary to conventional wisdom, Japan is not well positioned to develop close links with foreign countries' basic science. Our data suggest that the Dutch are likely to be much better at it.

THE PROPERTIES OF BASIC RESEARCH-PRODUCING INSTITUTIONS

In addition to the links between science and application, we need a better understanding of the properties of basic research-producing institutions, particularly universities and university departments. Public policy in the United Kingdom (and perhaps in other countries) increasingly assumes that there are advantages to greater scale and concentration in basic research activities, although there is no systematic evidence that this is the case. Similarly, there is talk of the need for choice and radical restructuring, without any clear idea as to how it will be done.

In this context, policy would be better informed as the result of research programme that combines recent advances in bibliometric methods, with accumulated experience in industrial economics in understanding the links between technological activities, firm size and industrial concentration. There would no doubt be similar room for debate over the adequacy of the various measures used, but similarly useful results would probably emerge, showing considerable variations amongst scientific fields in concentration and economies of scale.

What are the Unexploited Economies of Scale in British Basic Research?

Partly as what they would consider legitimate acts of academic self-defence, British scholars have been amongst the first to identify the problems to be clarified. As Hare and Wyatt (1988) have recently pointed out, little systematic evidence is available on economies of scale in basic research. In the United States, Frame and Narin (1976) found no economies of scale in biomedical research, when output was measured by numbers of publications. In the United Kingdom, there is now the possibility of making considerable progress in comparing the size distribution, degree of concentration, and productivity of basic research activities in different scientific fields, given recent improvements in the data on academic personnel, research grants, publications and citations in British universities.

In a preliminary analysis of 45 physics departments in British universities, my colleagues J. Skea and D. Hicks (1989) have come to similar conclusions to Frame and Narin: no unexploited economies of scale, when output is measured in numbers of publications. Before we can come to firm conclusions, the analysis will need to be extended to other scientific fields, and down into sub-fields. It will also be necessary to test the sensitivity of results to various measures of inputs and outputs: for example, McAllister and Narin (1983) found no economies of scale in US biomedical research in terms of the number of publications, but they did find higher citation rates amongst the larger institutions. In the meantime, a rigorous empirical case for greater selectivity and concentration in British basic research remains to be made.

How Do Basic Research Institutions Evolve?

Just as in the analysis of firms' technological activities, cross-sectional comparisons of size, concentration and efficiency – while useful – will also raise further important questions for theory and policy: in particular, how and why do the existing patterns come about? This leads on to four further questions, each of which is also central to the analysis of the dynamics of technical change, concentration and efficiency in industry:

1. Are large and productive research institutions good because they are big, or big because they are good?
2. What are the characteristics of productive institutions? To what extent do they grow out of accumulated scientific and managerial skills?
3. What is the appropriate organisational unit in which such skills are accumulated? Preliminary analysis by J. Platt (1988) suggests that it is not at the level of a university as a whole, but (if at all) in closely related subjects.
4. What are the mechanisms through which good research practice and productivity are diffused (or not) throughout the research community?

CONCLUSIONS FOR POLICY

Gaps in Knowledge

Conclusions for policy are bound to be tentative, given the still shaky theoretical and empirical base, which is why I have signalled throughout the paper where further research is required. The three most important subjects (in my view) are:

1. The economic and social usefulness of 'unstrategic' science, including the development of useful skills, and unplanned applications.
2. The sources of basic research skills and knowledge for UK firms compared to firms in other countries, particularly West Germany and Japan.

3. The structure, efficiency and dynamics of the British system of basic research.

Is British Science Particularly Useful to Britain?

In the meantime, one strong conclusion is that British basic science remains particularly useful to Britain. Arguments that this is not the case are based on mistaken assumptions. It is not the case that basic research is a universally available 'free good': assimilation and reduction to practice are costly. Nor is it the case that the international technological activities of large firms have eliminated the close coupling between British science and British technology. The main problem in using British science lies elsewhere: inadequate investment in R&D to take advantage of basic discoveries, particularly in the British engineering – mechanical, electrical and electronic – industries (see Patel and Pavitt, 1989a).

Should There Be More Selectivity and Concentration?

Another strong conclusion is that a convincing case for greater selectivity and concentration has yet to be made. If the US experience is any guide, technologists are interested in academic research, not just in the relatively few fields with direct applications, but also in the much broader range of fields where skills learnt through doing basic research are essential inputs to solving applied problems. This fact, together with the continuing importance of unplanned applications of basic science, argues for support for a diversity of fields in basic research. A similar case can be made for diversity of institutional support, given the lack of evidence of unexploited economies of scale in the performance of basic research.

Paying for Strategic Research

The above analysis also suggests that, as in the Science and Engineering Research Council (SERC)'s Biotechnology Directorate (Sharp and Senker, 1988), 'strategic research' should be funded jointly by governments and firms, for two reasons. First, firms must build their own competence if they are to assimilate successfully any results from such research. Second, participating firms can appropriate at least some of the benefits, since – in seeking potential applications – they will have learned how to combine the results of strategic research with other firm-specific assets, and this cannot be imitated overnight.

A Revised Case for Public Subsidy

This argument suggests that the justification for public subsidy, in terms of complete inappropriability of immediately applicable knowledge, is a weak one. In fact, the results of basic research are rarely immediately applicable, and making them so also increases their appropriability, since they become less general and less codified. In three other dimensions the case for public subsidy is stronger.

The first is training in research skills, where private firms cannot fully benefit from providing it when researchers, once trained, can and do move elsewhere. There is, in addition, the important insight of P. Dasgupta (1987) that, since the results of basic research are public and those of applied research and development often are not, training through basic research enables more informed choices and recruitment into the technological research community.

The second justification was originally stressed strongly by Nelson (1959), but has been neglected since then: namely, the considerable uncertainties before the event in knowing if and where the results of basic research might be applied. We now know from transaction cost theory that high uncertainty is one reason why markets are not necessarily efficient (Williamson, 1975). The probabilities of application will be greater with an open and flexible interface between basic research and application, which implies public subsidy for the former. The case for such a subsidy is strongest for 'unstrategic' fields of curiosity-driven research, the application of which cannot be foreseen.

A third, and potentially new, justification grows out of internationalisation of the technological activities of large firms, discussed earlier. Facilities for basic research and training can be considered as an increasingly important part of the infrastructure for downstream technological and production activities. Countries may therefore decide to subsidise them, in order to attract foreign firms or even to retain national ones. Recent interest in so-called 'science parks' might sometimes be one manifestation of this trend. Clearly there are dangers of competitive subsidy, the implications of which should keep game and trade theorists busy for some time.

A FOOTNOTE ON THE EFFECTS OF ANALYSIS ON SCIENCE POLICY

As an extended footnote to this paper, I want to warn against the dangers of expecting too much in the management of science – either for good, or for ill – from policy analysis by economists, bibliometricians and others. Both expectations and fears are probably exaggerated. It can be argued that the practical impact of this particular strand of academic analysis will be no different in kind from the impact of academic natural science on technology: it will help practitioners cope a little better with complex problems.

Thus one major influence of science on technology has always been through the development of techniques of measurement. From the telescope, through the cathode-ray tube, to instruments for controlling continuous production processes, measurement techniques pioneered by academics have found widespread practical applications. In the same way, cheap information technology has now opened the way for the diffusion of the methods of categorising, counting and comparing scientific papers, patent documents and their citation patterns that were first developed by academics more than 20 years ago, and have been in a continuous process of refinement since then. However, as with earlier techniques, successful assimilation into existing practice must first overcome three sets of difficulties.

The first is the resistance of established practitioners. Early attempts to measure and understand production activities in factories ran into the hostility of workers with craft and production skills, who feared that top management would thereby undermine the tacit knowledge on which their value was based. Similar hostility can be detected from those responsible for established systems for evaluating research, who fear that their power and influence will be undermined.

The second difficulty is that new measurement techniques often go through exaggerated cycles of optimism, followed by pessimism, about their eventual utility. There were, for example, the high expectations at the beginning of the 1960s about applications of the laser, followed by the pessimism at the beginning of the 1970s, before the widespread applications of the 1980s. Similarly, science and technology policy-makers are now finding out that many of the available data bases on scientific papers, patents and their citations have a two-year lag, and are therefore more useful for evaluating past performance than for the monitoring of current performance. There is also the danger of concentrating on what is easily measurable (e.g. numbers of patents and scientific papers; citations) neglecting what is otherwise important (e.g. the output and employment of graduate and post-graduates in science and engineering; the long-term and roundabout impact of curiosity-oriented research).

The third set of difficulties in applying new measurement techniques arise from the need to improve complementary techniques. We know that many applications of fast-moving electronics technologies depend critically for their success on improvements in complementary mechanical technologies. The effective use of bibliometric techniques often needs to go hand in hand with other improvements in refereeing procedures. Considerable improvements can be made through relatively simple applications of information technology: for example, the widening and deepening of accessible information on possible referees for research applications; ensuring that information is quickly available on applicants' past performance. And perhaps the arrival of FAX technology will make the UK Research Councils more willing and able to consult foreign experts as part of their normal refereeing procedures.

Finally, as we have seen, science influences technology, not just through techniques of measurement, but also through the background knowledge that it provides, and that influences both the identification of problems and ways in which they are tackled. Similarly, Luukkonen-Gronow (1989) has shown that science policy research is conceptual as well as instrumental in its impact. Conceptual influences may come from the provision of better data, such as international comparisons of scientific or technological activities. They may also come from better understanding of the factors influencing the quality and impact of scientific activities of the kind that would, it is hoped, emerge from the research programme I have suggested above. Conceptual influences on the policy process are inevitably diffuse and more difficult to identify than instrumental influences. For these reasons, they meet with less resistance from the established policy process; indeed the influences are insidious, and often not even recognised to exist. I shall be very happy indeed if that turns out to be the fate of this chapter.

REFERENCES

Arrow, K. (1962) 'Economic Welfare and the Allocation of Resources for Invention', in R. Nelson (ed.) *The Rate of Direction of Inventive Activity* (New Jersey: Princeton UP).

Averch, H. (1985) *A Strategic Analysis of Science and Technology Policy* (Baltimore: Johns Hopkins Press).

Averch, H. (1989) 'Exploring the Cost-Efficiency of Basic Research funding in Chemistry'. *Research Policy*, 18, pp.165–72.

Carpenter, M. (1983) 'Patent Citations as Indicators of Scientific and Technological Linkages', *Computer Horizons Inc.*, New Jersey.

Council of Science and Society (1989) *The Value of Useless Research: Supporting Science and Scholarship for the Long Run* (London).

Dasgupta, P. (1987) 'The Economic Theory of Technology Policy: an Introduction', in P. Dasgupta and P. Stoneman (eds), *Economic Policy and Technological Performance* (Cambridge: Cambridge UP).

David, P., D. Mowery and W. Steinmuller (1988) 'The Economic Analysis of Payoffs from Basic Research – (An Examination of the Case of Particle Physics Research', *CEPR Publication No. 122* Center for Economic Policy Research (Stanford University, California:).

Dertouzos, M., R. Lester and R. Solow (eds) (1989) *Made in America: Regaining the Productive Edge* (Cambridge, Mass.: MIT Press Mass.).

Eads, G. and R. Nelson (1971) 'Governmental Support of Advanced Civilian Technology: Power Reactors and the Supersonic Transport', *Public Policy,* 19, pp.405–28.

Fagerberg, J. (1987) 'A Technology Gap Approach to Why Growth Rates Differ', *Research Policy*, 16, pp.87–99.

Fagerberg, J. (1988) 'International Competitiveness', *Economic Journal*, 98, pp.355–74.

Frame, J. and Narin, F. 'NIH Funding and Biomedical Publication Output', *Federation Proceedings*, Vol 35, pp.2529–2532

Freeman, C., J. Clark and L. Soete (1982) *Unemployment and Technical Innovation* (London: Pinter).

Gibbons, M. and R. Johnston (1974) 'The Roles of Science in Technological Innovation', *Research Policy*, 3, pp.220–42.

Hare, P. and G. Wyatt (1988) 'Modelling the Determination of Research Output in British Universities', *Research Policy*, 17, pp.315–28.

Illinois Institute of Technology Research Institute (1969) *Report on Project TRACES* (Washington: National Science Foundation).

Irvine, J. and B. Martin (1980) 'The Economic Effects of Big Science: the Case of Radio-Astronomy', in *Proceedings of the International Colloquium onh Economic Effects of Space and Other advanced Technologies*, Ref. ESA SP 151 (Paris: European Space Agency).

Irvine, J., B. Martin and N. Minchin (1986) 'Is Britain Spending Enough on Science?', *Nature*, 323, pp. 591–4.

Isard, P., B. Martin and J. Irvine (1989) 'Trends in UK Government Expenditure on Academic and Related Research: Preliminary Results from an International Comparison with France, Japan, the Netherlands,

United States and West Germany' (mimeo), Science Policy Research Unit, University of Sussex, Brighton.
Jewkes, J. (1972) *Government and High Technology,* Institute of Economic Affairs, Occasional Paper 37 (London).
Kay, J. and C. Llewellyn Smith (1985) 'Science Policy and Public Spending', *Fiscal Studies*, 6, pp.14–23.
Kodama, F. (1989) 'Some Analysis on Recent Changes in Japanese Supply and Employment Pattern of Engineers' (mimeo), National Institute of Science and Technology Policy, Tokyo.
Luukkonen-Gronow, T. (1989) 'The Impact of Evaluation Data on Policy Determination', in *The Evaluation of Scientific Research*, (Ciba Foundation Conference: Wiley Chichester) pp.234–46.
McAllister, P.and F. Narin (1983) 'Characterization of the Research papers of US Medical Schools', *Journal of the American Scoeity for Information Science*, 34, pp.123–31.
Martin, B. and J. Irvine (1989) *Research Foresight: Priority-Setting in Science* (London: Pinter).
Mansfield, E. (1989) 'The Social Rate of Return from Academic Research' (mimeo), University of Pennsylvania, Philadelphia.
Mowery, D. (1983) 'Economic Theory and Government Technology Policy'. *Policy Sciences*, 16, pp. 27–43.
Mowery, D. and N. Rosenberg (1989) *Technology and the Pursuit of Economic Growth* (Cambridge: Cambridge UP).
Narin, F. and F. Noma (1985) 'Is Technology becoming Science?', *Scientometrics*, 7, pp.369–81.
Narin, F. and J. Frame (1989) 'The Growth of Japanese Science and Technology', *Science*, 245, pp.600–4.
National Science Board (1987) *Science and Engineering Indicators – 1987* (Washington).
Nelson, R. (1959) 'The Simple Economics of Basic Scientific Research', *Journal of Political Economy*, 67, pp.297–306.
Nelson, R. (1962) 'The Link between Science and Invention: the Case of the Transistor', in R. Nelson (ed.), *The Rate and Direction of Inventive Activity* New Jersey: Princeton UP).
Nelson, R. (1987) *Understanding Technical Change as an Evolutionary Process* (Amsterdam: North-Holland).
Nelson, R. and R. Levin (1986) 'The Influence of Science, University Research and Technical Societies on Industrial R & D and Technical Advance', *Policy Discussion Paper Series Number 3*, Research Program on Technological Change (Newhaven, Connecticut: Yale University Press).
OECD (1980) *Technical change and Economic Policy (Paris).*
Patel, P. and K. Pavitt (1989a) 'The Technological Activities of the UK: a Fresh Look', pp.113–54 in A. Silberston (ed.), *Technology and Economic Progress* (London: Macmillan).
Patel, K. and K. Pavitt (1989b) 'Do Large Firms Control the World's Technology?', (mimeo), Science Policy Research Unit, University of Sussex, Brighton.
Pavitt, K. (1987) 'The Objectives of Technology Policy', *Science and Public Policy*, 14, pp.182–8.

Platt, J. (1988) 'Research Policy in British Higher Education and Its Sociological Assumptions', *Sociology*, 22, pp.513–29.
Price, D. de Solla (1963) *Little Science, Big Science* (New York: Columbia UP).
Rosenberg, N. (1976) 'Karl Marx on the Economic Role of Science', in *Perspectives on Technology*, pp.126–38 (Cambridge: Cambridge UP).
Rosenberg, N. (1985) 'The Commercial Exploitation of Science by American Industry', in K. Clark, A. Hayes and C. Lorenz (eds), *The Uneasy Alliance: Managing the Productivity–Technology Dilemma* (Boston: Harvard Business School Press).
Rosenberg, N. (1990) 'Why do Firms do Basic Research (with Their Own Money)?', *Research Policy* pp.165–174. Vol 19, No 2,
Sharp, M. and J. Senker (1988) 'Promoting Bio-Technology: Eight Years of the SERC's Biotechnology Directorate', *Biofutur*, No. 74, pp.47–9.
Skea, J. and D. Hicks (1989) 'Scale Effects on the Output from UK Physics Departments' (mimeo), Science Policy Research Unit, University of Sussex, Brighton.
Soete, L., B. Verspagen, P. Patel and K. Pavitt (1989) *Recent Comparative Trends in Technology Indicators in the OECD Area*, International Seminar on Science, Technology and Economic Growth, OECD (Paris).
Stoneman, P. (1987) *The Economic Analysis of Technology Policy* (Oxford: Oxford UP).
Stoneman, P and J. Vickers (1988) 'The Assessment: the Economics of Technology Policy'. *Oxford Review of Economic Policy*, 4, pp.i–xvi.
Williamson O. (1975) *Markets and Hierarchies: Analysis and Antitrust Implications* (New York: Free Press).
Ziman, J. (1987) *Science in a "Steady State" the Research System in Transition*, SPSG Concept Paper No. 1, Science Policy Support Group (London).

3 Are Some Science Policy Issues Inevitable, Irresolvable and Permanent?

FREDERICK DAINTON

THE KINDS OF SCIENTIFIC ACTIVITY

Together with three other chemists from Cambridge, I was sent in June 1939 to the Royal Aircraft Establishment, Farnborough under a government scheme which was activated by the imminence of war. One problem I was assigned was to find some chemical vapour which, when mixed with the hydrogen used to inflate barrage balloons (shortly to become a very visible feature in the defence of our cities against enemy bombers), would prevent that gas igniting when the balloon was punctured by a high-explosive incendiary bullet. Because my forthcoming PhD thesis was focused on the nature of explosions in mixtures of hydrogen with oxygen, my combined knowledge of this and of flame propagation, on which a contemporary student was working, was sufficient for me to assert with confidence that no such magic ingredient could exist. An engineer with whom I discussed this matter disagreed on the grounds that 'every question has an answer' and I was prevailed upon to embark on experimentation which, predictably, was quite fruitless.

About half-way through the war I was presented with another practical problem, namely, that aircraft in the fuel tanks of which an enemy bullet containing phosphorus had become lodged, would spontaneously inflame some time after landing at the North African airstrip. A quick perusal of the literature concerning the inflammation limits of white phosphorus and a few confirmatory experiments, together with a short investigation of the effect of aviation fuel vapours on these limits, soon enabled a procedure to be devised which prevented the unwanted conflagration. Some years later, when I was researching more thoroughly the oxidation of phosphorus, I found the essential qualitative information was discovered empirically by Robert Boyle some 250 years earlier when he was allowing his curiosity to drive him to investigate what was then called the icy 'Noctiluca', a demonstration of which to the Court was an impressive entertainment.

There was not much time to think during the war, but a few years later I realised that these events illustrate some points which must never be forgotton by science policy-makers. With hindsight I now know that my riposte to the engineer should have been 'What is the question?' to which I expect he would have answered, 'What chemical compound can in small amounts prevent the conflagration of hydrogen-filled balloons struck by an incendiary bullet?' I could then have answered his question correctly with the words, 'There cannot be such a compound', and done so on the basis of existing knowledge which had been acquired with no practical end in view. To provide that answer also required no further experiment and therefore there was no cost. The second case I described differs from this only in the minor respect that the answer to the practical question required a slight extension of existing knowledge which, predictably, could be obtained by existing methods and at an ascertainable cost. In both cases no theoretical advance was necessary in order to answer the practical question; the necessary theory already existed and was available simply because the curiosity of some gifted individuals had been excited by the fact that some chemical reactions, and especially gaseous oxidations, can proceed quietly and yet, with only a slight change of external conditions, become explosive. By a strange quirk of fate I have for another purpose, just written a biographical memoir of the man, N.N. Semenov, who provided the essential theory of this phenomenon as a part of the theory of chain reactions, and it is quite clear that what motivated his work was the desire to devise a theory which would accomodate already known attested chemical facts, bringing them into a previously unattainable rational relationship, on the basis of which testable predictions about the unknown could be made. Semenov did what many academic scientists do: he saw a *problem* which excited his curiosity because he felt it must have a solution, and he was driven forward by that curiosity and what increasingly appears to be the power and elegance of the explanation he was developing, and he did so without any pressure as to deadline; indeed such pressures would probably have been counter-productive.

I hope these examples help to define what goes on at the extreme ends of the range of activities in which scientists, by which I mean scientists, engineers and technologists, engage. At one end there are scientists trying to find the cheapest and best answers to practical *questions* posed by their employers and customers, and to do so within the shortest possible time. At the outset of such work it is known that the answer, which may result in a new process or product or design, is in principle attainable and by methods involving little conceptual novelty and likely to yield surprise-free results. Some 18 years ago I christened this as *tactical science*.[1] Many other actions may of course need to be taken in what is pre-commercial *development* work, before the product can be manufactured and the process optimised and either the product or the process put on the market.

At the other end of the range is *basic, self-chosen, curiosity-oriented* (many other words have been used) *science*, rarely undertaken to answer a practical *question* though, as I have tried to illustrate, such investigations very frequently in later years facilitate the identification of the means to answer as yet unforeseen questions. The main motive to undertake this *basic* scientific work is *to investigate a problem,* to find an explanation, to extend, modify or replace an idea or a theory; substituting something else of greater comprehensive and predictive power. By its very nature this kind of work is speculative and new thought or experimental methods may have to be devised to explore the problem. By definition, no time frame can be set for the attainment of a successful solution; indeed the solution may prove to be wholly elusive or at best partial.

Between these extremes lies *strategic* science. This is a broad field of enquiry identified as worthy of pursuit by some policy forum because, although the precise question, the answers to which would be valuable and beneficial, cannot at the time be formulated and the time at which useful answers will emerge is therefore still quite unknown, it is nevertheless felt that research in this area cannot fail to yield results of practical value sometime. The most well known British example of this is a decision of the MRC (Medical Research Council) to support work on the chemical structure and conformation of vital cellular constituents. Out of this grew the whole field of molecular biology, with consequential benefits to agriculture, medicine and veterinary practice, not to mention the creation of a new industry of biotechnology. But perhaps social scientists would be more interested in the example of the same Research Council's action in establishing a Laboratory of Applied Psychology at Cambridge, initially under Sir Frederick Bartlett's direction, which had very practical outcomes in, for example, selection of humans for specific tasks or further education, in workshop layout design, treatment of prisoners and so on.

As has often been demonstrated, such strategic work has the potential for making fundamental advances in knowledge, so that the boundary between it and basic science, with which it shares the element of timelessness, is not always sharp. Nor is the boundary between strategic and tactical research always a complete disjunction; in fact it is often better to regard scientific activity as a spectrum from basic to tactical. Let me quote another example from one of my own fields of basic research, the problem of the understanding of the effect of radiations from radionuclides or particle accelerators in bringing about chemical change. It was early realised that such radiations damage plants and animals and do so mainly through their effect on the major cellular component, water, and that exposure of humans would have deleterious effects on them. (We all remember that Marie Curie herself died from persistent aplastic anaemia undoubtedly caused by over-exposure to radiation.) But the other side of that coin is that such radiation, locally applied to diseased tissue, and especially tumours, could, by destroying it, be a form of beneficial therapy. The second impetus to this fundamental work came when the construction of heavy water-moderated nuclear reactors was being contemplated. Satisfactory answers to both the biological and the engineering questions cannot be forthcoming until certain fundamental problems of understanding of the radiation chemistry of water are resolved. Here one can detect a conflict of two time scales, the limited and the infinite.

There are also certain linkages between practical *questions* and fundamental scientific *problems*. Thus a need to answer a practical question, such as, 'What is the best medicine for a particular disease?', may strengthen motivation to investigate a problem, in this case that of the nature and mechanism of the action of chemicals on living organisms. Conversely, research to solve problems, even when undertaken without a practical end in view, will ultimately almost invariably make the answering of some practical question easier. As a well known example I would cite how an understanding of the effect of the magnetic properties of atomic nuclei on the energy levels of molecules in strong magnetic fields has led, not just to more powerful methods for the elucidation of chemical structure and mechanism, but also to non-invasive examination of the interior of human beings and the exploration of the metabolic processes occurring within them.

SCIENCE AND GOVERNMENT: SOME EARLY PROBLEMS AND QUESTIONS

As long as science was cheap and could be paid for by the investigator himself or by his patron, and its effect on society was small, science policy was not a term in use, let alone an issue for government. Nevertheless governments did from time to time invest in science when they saw a public advantage, for example in establishing the Board of Longitude (1714) or by supporting voyages of exploration and discovery. Now, however, science is expensive and it has powerful effects on the ways in which we live and work and even on social attitudes and personal relationships, as well as influencing the economic success or failure of individuals, enterprises and society. A host of questions come to mind. Where is science best done? Who is to pay for it? How much money should be invested in it? How is it to be managed? Whose intellectual property is it? How can it be exploited? And so on.

This is an impossibly long menu of questions for discussion in a single short presentation. For this reason I shall restrict myself to some questions of policy concerning publicly-funded science. These questions are more difficult than the science policy questions which arise within a firm where the overriding concern is that the scientific effort should contribute significantly, and to the expected extent, to the economic survival and commercial success of the enterprise. Private research cannot be entirely disconnected from publicly-funded research, simply because all scientists are trained in the public sector and, in addition, there are strong arguments for close links between the public and private sectors, to the benefit of both.

HOW GOVERNMENT ORGANISATION FOR SCIENCE EVOLVED

The government as paymaster works through several agencies. First, there are the Departments of State, most of which need science to help them shape policy and, sometimes, to put it into effect. Thus the Department of Defence needs operational research to help define some policies; it needs other kinds of research for the specification of the design of weapons and other equipment and their carriers on land and sea and in the air, as a preliminary to procurement. (In passing, it should be said that the percentage of research into weapon systems undertaken in-house by government is disadvantageously high in the United Kingdom, and that it would be better for government and manufacturers if the latter undertook more of the actual research and development.) On this basis, a Department of State would seem to be like a very large firm in its requirements for research and should manage it accordingly, either performing it in-house, in its own establishments, or out-house, by contract with suppliers which might be private or public (often higher educational institutions and establishments). Most of the questions I have posed are thus, in principle, quickly and simply answered. Sometimes, however, there are complications. Different departments may need similar scientific knowledge for different purposes and even a wide range of institutions and individuals outside government may require it as well. A good example is radiological protection information, needed by the Department of Health for the safety of medical personnel and by the Department of the Environment and the Ministry of Agriculture, Food and Fisheries to meet their responsibilities with regard to the risk of harm through over-exposure from radionuclides getting into the food chains; the Health and Safety Executive need the information for reasons of protection of personnel at work and for the licensing of reactor design; manufacturers who use radionuclides or accelerators need it. Even those responsible for the health of high altitude airline pilots, and individuals concerned about radon in their houses, also need radiological information. With so many customers there is a clear case for a free-standing body like the National Radiological Protection Board which can charge for its services and yet have some core funding, if necessary from more than one government department, to ensure that it can maintain research capability to solve *problems* and thereby place itself in a position to answer practical *questions* quickly.

Some 70 years ago the great Lord Haldane clearly identified the sort of research I have just described, which he considered should be 'supervised', by which he meant managed and paid for, by a department of government, because it was necessary for that department to carry out its own task. Haldane also characterised another category of research which he called 'research for general use'. This was the by-product of a Committee on the Machinery of Government chaired by Haldane, and the committee considered that research for general use should be supported by government as part of its business of serving the public with what he described as *research and information.*[2] Because it was for the general good he thought that it should be administered independently of any department and therefore free from any political and administrative pressures. Christopher Addison, a remarkable medical student and later Professor of Anatomy in the University of Sheffield and the first British Minister of Health, had earlier and quite independently come to the same conclusion. And so the *arm's-length principle,* as being applicable to 'science for general use', was codified if not born, and the idea of research councils to administer this research emerged, with the council members to be appointed by the Privy Council.

Since 1919, when there were only two research councils, the Department of Scientific and Industrial Research (DSIR) and the Medical Research Council (MRC), the Agriculture Research Council (ARC, now AFRC) was created in 1931; and following the Trend[3] and Heyworth[4] Reports in the 1960s, the DSIR was split, one portion becoming the Science Research Council (SRC, now SERC) and the Social Science Research Council (SSRC, now the ESRC) and the National Environmental Research Council (NERC) was created and all the Research Councils were transferred from the Privy Council to the Department of Education and Science. The DES set up a Council for Scientific Policy (CSP) to allocate resources to the Research Councils and advise the Secretary of State on his newly-acquired responsibilities. At the time there were a number of influential persons who were deeply concerned about the transfer from direct Treasury funding to Departmental funding and some were also against the creation of a Department of Education and Science to which responsibility for funding the universities, through the University Grants Committee (UGC), was also given. The objections were based on a perceived abrogation of the arm's length principle, although a few hankered after a Ministry of Science, which had been tried and failed, as has also proved subsequently to be the case in

Australia, simply because such a Minister may be accorded a trans-departmental co-ordinating role but is never given any money and therefore lacks 'clout' to modify effectively government science policy.

I always believed the decision to be correct because:

1. 'Science for general use' comprises by definition all *basic* research and a considerable part of *strategic* research, and the individual programmes are often closer intellectually to one another than their dimly perceived possible outcomes might be to the specific needs of any department
2. The best-quality basic and strategic research takes place when the researchers have a high degree of freedom. Significant ideas, advances in methodology and novel techniques are each more frequently the product of individual minds and can rarely, if ever, be programmed to appear by some administrative *fiat*, the product of a committee. Administration of basic and some strategic science is more a matter of recruiting able people and creating the conditions for them to give of their best, that is, it is best when policy is responsive rather than *dirigiste*. Again by definition these conditions are difficult to achieve within a Department of Government.
3. Universities, on the other hand, are by their nature, dedicated to free enquiry and they are also communities of researchers in many fields and therefore, in principle, it is easier to develop the synergism of collegiality and multidisciplinarity within universities.
4. Universities are the places where the cast majority of the apprentices in research, that is, graduate students, are trained and the quality of that manpower and the training it receives is of paramount importance to the success of the whole national scientific enterprise for many years to come.
5. Yet another reason for regarding the universities as the best locus for 'research for general use' is that undergraduates are enthused or repelled by a subject to some considerable degree, depending on its liveliness, which they often measure in terms of the opportunities for making significant and important advances. Clearly this judgement cannot easily be made if their teachers are not committed to active scholarly or research work. This was beautifully expressed by the Indian mystic and philosopher Rabindranath Tagore: The teacher who has come to the end of

his subject, who has no living traffic with his knowledge but merely repeats his lesson to his students can only load their minds; he cannot quicken them.

SCIENCE AND GOVERNMENT: SOME CURRENT QUESTIONS

For almost 25 years successive UK governments have been content to leave the structures of publicly-funded scientific research and development largely as I have described them: most tactical work to serve the statutory duties and the procurement or regulatory policies of government departments to be funded and controlled by the departments themselves, and most basic and much strategic work to be funded on a relatively loose rein through the Department of Education and Science via the Dual Support System, in which some monies reach the performers, mainly academic staff, through the UGC grant to the universities while others come largely via the Advisory Board for the Research Councils (ABRC), the successor to the CSP, and the five Research Councils (see Figure 3.1). There have, of course, been some minor changes. The need for a central council to formulate advice on national strategy has been felt and met first by the Central Council for Science and Technology (1967/70), then partially by the Central Policy Review Staff (1970/80); next the Advisory Council of Applied Research and Development (ACARD) was set up to balance the ABRC, and both were followed by the creation of the Advisory Council for Science and Technology (ACOST) placed at the centre, that is in the Cabinet Office. Under the aegis of the Chief Scientist in the Cabinet Office, statistics about scientific R&D have been greatly improved. Parallel with these developments has been that in the European Community, both through specific agencies such as the European Space Agency and through the initiatives of DG 12 of the European Commission which affect member countries including Britain. New UK bodies with specific missions have been created and initiatives such as the Alvey Programme taken. Of course, there are always arguments about the boundaries between the work of the Research Councils and that of Departments, and in 1972 this led to a transfer of funds from the ARC, NERC and MRC to some government departments, a process reversed nine years later for the MRC, because it achieved no purpose other than the creation of a bureaucracy.

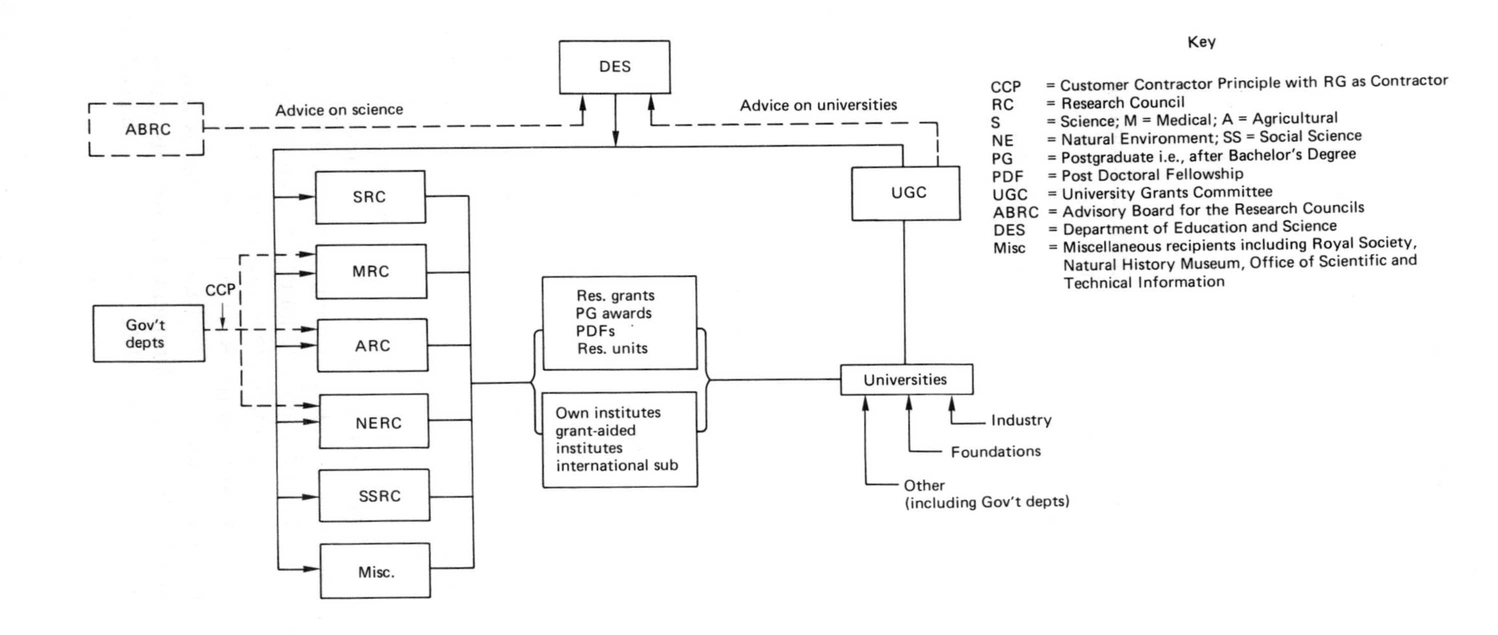

FIGURE 3.1 *The dual support system*
Note: On 1 April 1989 the UGC was replaced by the University Funding Committee (UFC), a statutory body.

But to the observer these changes seem to be but minor perturbations compared with what might happen if present trends continue unabated. The signs of rising concern are plain in the press, where science policy has received much critical attention, and it is also manifested in the worry of scientists, a worry which has led to the unprecedented establishment of the 'Save British Science' movement.

Universities are also alarmed by UGC/UFC funding which declines per capita in real terms, and which is applied through formulae lacking the requisite intellectual and statistical justification, and using criteria of judgement from which many professional bodies dissent. There was also an ill-judged attempt to impose patterns of activity on universities which would force them into categories – R for research, T for primarily teaching and X for some unknown intermediate category – all evidence of 'top-down' planning. Partly as a result of a greater insistence on accountability in the use of public funds and monitoring the whole process, management jargon has become all the rage and all funding agencies and universities are to have corporate plans, performance indicators, efficiency measures and so on. It would be easy to ridicule some of these things, but collectively they are felt to be onerous and threatening. It is pertinent to ask whether, as far as science policy is concerned, the critical point has been reached, if so what is the cause and how any perceived new and changed situation is to be managed or adapted to.

One of the characteristics of science is its auto-acceleratory character, like the branched chain reactions with which I opened. This is because all scientific advances always lead to the means of creating further advances more effectively and rapidly than in the past. At the same time scientific research costs money and continues to demand more money. Consequently there is a confrontation between the scientific enterprise, with its in-built characteristic of explosive growth which carries with it a demand for resources to sustain that growth, and governments, which before they part with any resources for which they have many claimants other than science, want to know what assurances they can have that the investment they are asked to make will be used to best effect, and yield commensurate benefits which outweigh those either of investing elsewhere or of giving the money back to the tax payer. At present government is not prepared to sustain that natural growth, and many non-scientists support the stance because, throughout their lives, they have witnessed *growth* in government support for scientific and technological research and development accompanied by industrial *decline*. Clearly we must now probe whether the implied inverse relationship between these two qualities is valid and this necessitates a look backwards over two centuries.

REALITY AND ILLUSION IN TWO CENTURIES

The Industrial Revolution, which was the means of the massive augmentation of the feeble power of the human biped by the power generated by the combustion of coal or wood, began in Britain in the late eighteenth century. It was the result of the work of independently minded, often religiously dissenting, persons such as the engineers James Watt and Matthew Boulton who were in frequent contact in the Lunar Society with kindred spirits like the doctor, Erasmus Darwin, the innovative manufacturers Abraham Darby and Josiah Wedgwood and the Yorkshire Unitarian Minister of revolutionary views, Joseph Priestley, who discovered oxygen and many other gases. Their education owed nothing to England's two ancient universities, but many of them were Fellows of the Royal Society. Half a century later, by his studies of the laws of electricity and electrochemistry, another Fellow of the Royal Society, Michael Faraday, was laying the foundation for the conversion of steam power to electrical power. There ensued the Electrical Revolution, based on the large-scale generation of electricity and its transmission for lighting and heating to every home and factory, which led to electric traction and ultimately to the almost instantaneous transmission of voice and visual image through cables or via radiation to any selected receiver.

With such a head start it is not surprising that Britain was the first European country to transform itself from a primarily rural and agrarian society to an urban mercantile and manufacturing one, thereby earning for itself the soubriquet of 'the workshop of the world'. Accompanying this change was the growth of military power and the acquisition of more overseas territories to form a vast empire. The Great Exhibition of 1851 must have seemed an entirely justifiable celebration of success based on British inventiveness and initiative in many fields. And yet in this very success lay the seeds of decline, so that by the end of the nineteenth century Germany's exports exceeded those of the United Kingdom in both value and volume, and Germany had a more securely science-based industry than Britain. For example, her chemical industry, initially founded on the synthetic dye industry which was, ironically, derived from the work of the English organic chemist W.H. Perkin Senior, far surpassed that of the United Kingdom in size and efficiency.

With hindsight the cause of Britain's decline can be discerned. The very fact of empire, with its cheaper raw materials and the captive markets, took away the incentive to maintain a competitive edge over other countries not so blessed (or cursed, if your will). Moreover, Germany had learned, largely from the tutelage of von Humboldt, the Prussian Minister for Education, the secret of rapid knowledge transfer between the knowledge creators and the knowledge exploiters. It was he, by his foundation of the University of Berlin, who, with the organic chemist Justus von Liebig, encouraged the system of dual appointments straddling industry and the universities, thus forming a very effective bridge for the two-way flow of ideas and people. Britain should have observed this and learnt the lesson but it was not until the First World War, that Britain learned the hard way from Germany's ability to survive a blockade through self-sufficiency in large measure based on the effective application of science. One British statesman, the redoubtable Haldane, had learned the appropriate lesson more than two decades earlier through the perspicacity of his canny Scots lawyer father, who sent him to study philosophy in Göttingen rather than Oxford, so as to avoid 'catching the idle habits of the Oxonian' and, as I have already indicated, Haldane worked constantly to ensure that the native intelligence of able Britons was cultivated in the universities and applied through science to benefit the nation.

The Second World War taught us the same lesson, but in a different way. Germany had disadvantaged herself by making life so intolerable for able and especially Jewish scientists that many emigrated before the war, whilst Britain's scientists were used in the war effort. University science staff invented operational research, penicillin and the magnetron valve, so essential for radar, while those in industry had already developed new substances, such as polythene, perspex and terylene, to name but a few. Then, after the war, we fell into a trap of our own making. An illusion was propagated. Using the above, and other examples influential scientists claimed that government investment in science was bound to bring national dividends. 'What is good for science is good for Britain' might have been their slogan, and a grateful people and government, once the immediate postwar austerity had been ameliorated, showered them with gold so that, for a significant period, the percentage of the GDP spent on science in Britain was amongst the highest anywhere. I well remember that, in the heady days of the 'white-hot technological revolution', we members of the Council of Scientific Policy did not regard an annual growth rate of the science budget *in real terms* of 10 per cent as in any way abnormal.

The scientists, myself included, had a bonanza, which is not to say that we were idle. We were industrious and our work was regarded by our peers in the rest of the world as highly meritorious. Figures 3.2 and 3.3 illustrate how it was cited by them and how we performed

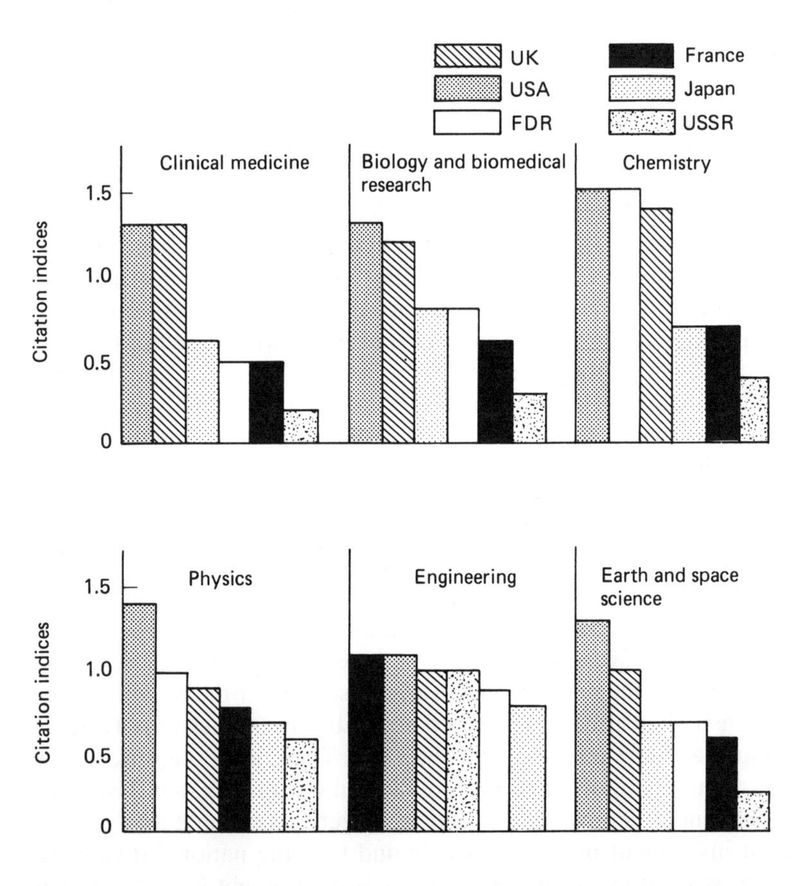

FIGURE 3.2 *Citation indices for six countries' publications in medicine, biology, chemistry, physics, engineering and Earth and space science, 1973*

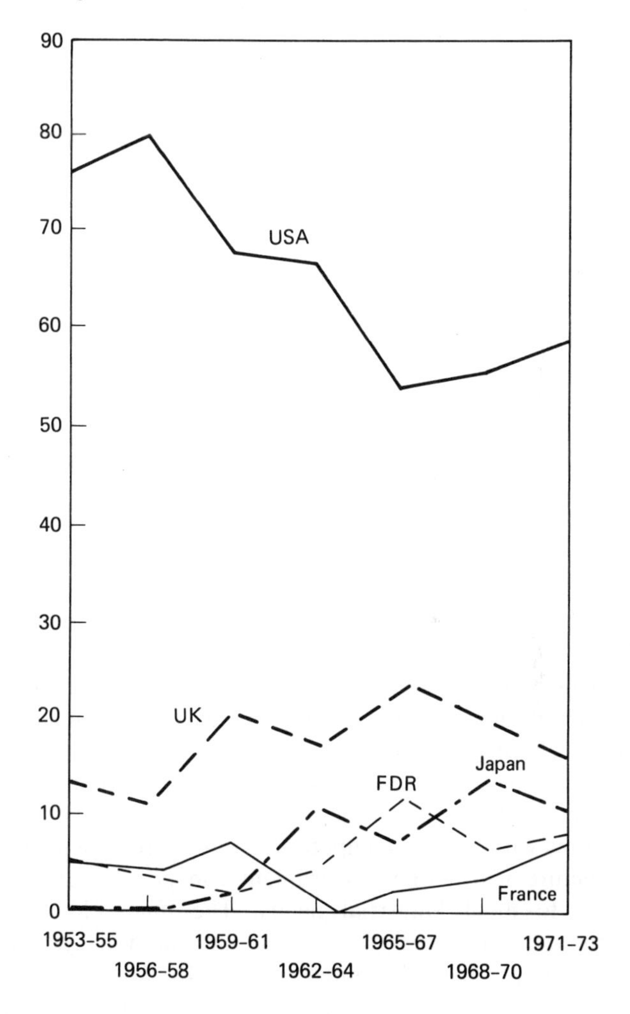

FIGURE 3.3 *Major technological innovations in UK, USA, France, FDR and Japan, 1953–73*

creatively in generating radical innovation. But the country's economic and industrial decline relative to comparable countries continued. When, following the world recession beginning in the mid-1970s, the day of economic reckoning arrived for Britain, the scientific community and the science vote were vulnerable on two grounds, one justified, the other invalid. The justifiable argument was that science should take its share of what the government deemed to be necessary cuts in public expenditure. The invalid, but one which I am convinced had strong emotional appeal to many ordinary citizens, was that 'the government has increased expenditure on science and the country's economic performance has deteriorated; therefore that expenditure should be reduced'. The obvious non-sequitur in this statement needs no comment, but the fact that it was made at all emphasises the need for better public understanding of the relationship, if any, between national economic success and government investment in science and whether further changes in government policy are necessary. So let me conclude by discussing what should be the government involvement *in tactical, basic and strategic science,* in that order.

HOW SHOULD GOVERNMENT BE INVOLVED IN SCIENCE?

My starting point is that wealth or added value is created not by governments but by industries which make a product or a process and sell it successfully and, in Britain's case, export much of it successfully. There are no obvious roles for government in this process, other than to ensure that British science-based industry competes on equal terms with foreign competitors, and to create and sustain that infrastructure which only government can provide and at a level adequate to meet industrial needs. The main elements of the infrastructure are access to a world information system and an adequate supply of well-trained people. In Britain the former is satisfactory, the latter is not. As an example of this I would cite the fact that the percentage of the labour force having qualified in properly organised apprenticeship schemes in Germany in 1978 was 59.9 per cent whilst in Britain it was 30 per cent and the corresponding figures for graduates were 7.1 per cent and 5.5 per cent respectively. Moreover a steady decrease in the percentage of school leavers opting for mathematics and science, to which I first drew attention 21 years ago[5] has not yet been reversed and this, combined with the one-third decrease in the size of the 21-year age group between 1983 and 1995, means that an acute shortage of graduates in the physical sciences and engineering is inevitable during the remainder of this century. Industry, being already aware of this, will use the obvious method to maintain recruitment, with damaging effects on science education at all levels, for which a high price will later be paid. This problem requires urgent attention by government and is not soluble by market forces unless government is prepared to pay market rates for science teachers. In addition, reducing the quality and range of university research, or making it unduly dependent on specific contract work, will constitute a depletion of the country's scientific capital.

All governments, observing that the country's share of a particular international market is diminishing, are tempted to intervene to rectify the position and even the most free market ideologues from time to time succumb to such temptation. The obvious point of intervention in highly science-based industries is to award money to them for research or for raising awareness levels or even acting as an additional customer for research done by them. But, as Sir Robin Nicholson has pointed out when he was Chief Scientist in the Cabinet Office,[6] scientific research and development paid for by an individual firm is likely to be better managed and to have a more positive effect on market share than any government subsidy. The contrast between the United Kingdom pharmaceutical industry, which in the period from 1975 to 1984 received no government money, and the electronics industry which did, is illustrated in Table 3.1.

There is one area of tactical science where concern for the citizens' welfare may compel government support for strategic and basic science. This is work to ensure that *regulations* are soundly based to enable standards to be maintained, for example, in foods, drugs, safety at work, radiological protection, water purity and sewage disposal. This is so obvious that I shall nto consider it further.

I turn now to basic science, which is of vital importance to strategic science as well as to tactical science of the future. Because of its heavily rising cost and its relative unpredictability of outcome by a specific time, it is very unlikely to receive adequate philanthropic funding from the private or corporate donors, although some trusts and foundations are still active in this field. Government has to be the major financial backer and, as I have explained, its agent in this matter is the Department of Education and Science. This Depart-

TABLE 3.1 *Comparison of the percentages of industrially-performed R&D funded by UK industry or government with the performance of those industries*

Electrical and electronics sector				
(i) R&D funds (£m) spent by industry	658	451	731	
government	345	499	912	
in year	1969	1975	1981	
(ii) Exports	1431	3059	3871	6969
Imports	1429	2931	4683	9798
Exports as % of all OECD	8.0	7.5	7.0	6.5
Year	1974	1978	1981	1984
Pharmaceuticals				
(i) R&D spent (£m)	30	79	251	490
As % gross output	6.6	7.3	11.4	11.7
Year	1970	1975	1980	1984
(ii) Exports	373	654	852	1222
Imports	97	201	298	542
Exports as % of all OECD	12.6	12.0	12.6	12.3
Year	1975	1978	1981	1984

Source: Nicholson (1988).

ment makes grants to universities and polytechnics, thereby providing most of the salaries of the academic staff, a large fraction of whom are also engaged in basic research, and the DES also contributes most of the direct library and laboratory costs, though to a diminishing extent in recent years. Secondly, through the other arm of the now creaking dual support system, the Department supports postgraduate research students, postdoctoral Fellows, research units and awards grants for scientific hardware and for employment of research and technical assistance.

During the 1980s a crisis developed in basic research, owing to the collision between the inherently auto-acceleratory growth of science, and the corresponding demand for money, with what were, in real terms at best level, and using realistic deflators, the diminishing funds government was prepared to give. One obvious consequence is a dissatisfied scientific community, of which the 'Save British Science' campaign is one manifestation, and a more important one is that bodies like the ABRC and the ACOST, exercised by how best to deploy their limited resources and subject to constant pressure for accountability for the funds they deploy, concentrate on policy and organisation. The latter has good and bad consequences.

A good consequence has been to tackle the question as to whether there should be five Research councils or a unified Research Council. As I tried unsuccessfully to bring this about as long ago as November 1970, when the external financial pressure was smaller, my position will be clear. I welcome the Morris Report and I hope that it or something like it becomes a reality.[8] I do so firstly because basic science is in fact a seamless robe. Developments in any area frequently enable advances to be made in a quite different field (as witness the effect of physics on chemistry, biology, earth sciences and medicine), and new opportunities and problems often arise which can only be tackled by multidisciplinary approaches. In fact boundaries between subjects have never had a sure philosophical foundation and now progress in science often demands their removal. Such fluidity of need and opportunity requires for success a flexible organisation and the minimum of 'baronies', one which is able to redeploy human and material resources within a fixed envelope in response to need. And in principle the percentage of income spent on bureaucracy should diminish.

The bad consequences outnumber the good. All science administrators know that when money is tight selectivity has to be greater. Added to this Research Councils are now enjoined to have a policy expressed in a so-called 'corporate plan'. By being 'purposeful', which actually means being more *dirigiste,* they hope to attract not only 'brownie points' but also pounds sterling from their paymasters. So the balance in funding steadily shifts away from basic science towards strategic and tactical science.

Such top-down management has various effects. The psychological effect on applicants for funds is to deflect them from self-chosen, that is basic, work to that which somebody, or group of somebodies, higher up has chosen and approved. And yet we know that, unless gifted young minds are able to have some freedom and opportunity to follow their own ideas, the seed corn of potential future harvests will not germinate. Concurrently with these financial circumstances there exists a currency in buzz words such as *centres of excellence* and *critical mass,* whose applicability to different areas has not been carefully examined but which are used to urge the scientific community to accept *concentration* of resources and effort. I can fully understand the need to concentrate physical equipment within particular institutions where it can be well serviced and maintained, but I do *not* see that it follows that would-be workers in that field attached to other institutions should be excluded from the use of it. By all means let us concentrate equipment, but we do not need to concentrate staff to the extent of totally depleting some institutions of staff in certain subjects; that course is an absolutely reliable way to ensure that, in unfavoured institutions, that subject will disappear, and even if it survives in the curriculum the nature of its presentation will be, to use Tagore's words, only to 'load, not quicken the minds of the young'.

Top-down planning has spread by contagion to the other arm of the dual support system. We have a new doctrine that any university department in physics of chemistry with less than 20 staff should disappear. This is obviously pedagogic nonsence, but since I have dealt with that point elsewhere[9] I will restrict myself here to commenting that, since at least half our universities have a neighbouring university within approximately 30–45 minutes' driving time, and even more have a polytechnic much closer to hand, any alleged deficiencies of teaching of small departments can easily be overcome by co-operation in teaching between neighbouring institutions. It is surprising that neither the planners nor the institutions themselves have moved in this direction, given the long and successful history of Britain's participation in international scientific collaboration in nuclear physics, astronomy, high-flux neutron beams, synchroton radiation and so on.

Also insiduously filtering down, even to basic research, are management techniques. Scientists are to have performance indicators. It is doubtful whether these are appropriate, reliable or effective because any scientist can learn to manipulate such crude measurement methods. The real danger is that these procedures will deflect able people from science or distract them from following their own noses in research. Raman has told us that he was impelled to study the diffusion of light by liquids after seeing the blue opalescence of the Mediterranean. I fear that, in modern times, to have ventured from this field into fluorescence studies, which led him to discover the Raman Effect, would have led to an early cancellation of his grant!

Faced with this situation the scientific community seems to me to have done little. Learned Societies have been relatively silent, perhaps because some of their luminaries, in the interest of protecting their own work, became collaborators with the process and so the community was divided. Of course, all scientists are agreed that they want more money, but that plea will butter no parsnips. They often make claims which are at best weak: for example, that high technology spins off into industry from technically advanced basic scientific projects, whereas Lew Branscomb, recently retired Vice-President for research in IBM, who has studied this matter carefully, tells us that case-studies show that this is not so, but that adoption by industry of high technology occurs by a 'trickle up' method.[10] And British scientists have failed to emphasise that it is not any deficiency in British science and inventiveness which is reponsible for poor industrial performance in industry, because the United Kingdom is one of only three OECD countries which, in the years 1979–86, had a favourable balance of payments in technology; that is, the value of the technical improvements, processes and products being developed here and then exported exceeds that of those we imported. The fault of poor performance in some science-based industries must lie elsewhere and ultimately, presumably, in the boardroom, perhaps in a failure to invest. Implementation of the recommendations of the Accounting Standards Committee that R&D expenditure should always be given in the annual accounts of companies could, by highlighting this problem, contribute to its solution.

CONCLUSIONS

So what do I judge to be the proper course for government? In a sentence, government should divest itself of some of its existing funding of industrial research and use the resources thus released to enhance the support which it gives to the dual support system. Within that system it should reorganise the Research Councils into a new National Research Council and take a hard look at whether it is not being over-interventionist, especially in basic science. At the same time, government should acknowledge and take vigorous action to remedy the difficulties faced by science, engineering and technology education in schools and further and higher education, and provide the supplementary funds which are so urgently needed to correct imbalances and mismatch with manifest future manpower needs. I do not pretend that this will provide a complete solution to some science policy problems, and in particular I believe that there is an inevitable permanence in the particular problem of the intrinsic expansionist force of science itself encountering the immovable object of control of government expenditure.

NOTES

1 *The Future of the Research Council System*, report of a CSP Working Group under the chairmanship of Sir Frederick Dainton (London:HMSO, Cmnd 4814, 1971)pp.3–4.
2. *Report of the Committee on the Machinery of Government* (London:HMSO. Cmnd 9280, 1918).
3. *Report of the Committee of Enquiry into the Organisation of Civil Science,* the Trend Report (London:HMSO, Cmnd 2171, 1963).
4. *Report of the Committee on Social Studies*, the Heyworth Report (London:HMSO Cmnd 2660, 1965).
5. *Enquiry into the Flow of Candidates in Science and Technology into Higher Education* (London:HMSO, Cmnd 3541, 1968).
6. R. Nicholson, 'Research investment: the key to a successful science and engineering-based industry', *Science and Public Affairs,* III (1988). 28–36.
7. Review of the Research Council' responsibilities for the biological sciences', report of a committee under the chairmanship of J.R.S. Morris, CBE
8. Dainton. 'Demise by size?', *Chemistry in Britain,* XXV (1989)pp.781–872.
9. Lewis M Branscomb, lecture to the American Academy of Arts and Sciences, 13 April 1988, communicated privately to the author.

4 The Management of Science in the 1990s: An American Perspective

ROBERTA BALSTAD MILLER

Given the difficulty in obtaining funding sufficient to support the escalating costs of scientific research in both the United States and the United Kingdom, an enquiry into the management of science is both timely and useful. It is a subject that deserves wide discussion, not only among scientists, but also by those involved in setting national priorities and policies and by the lay public. For, although the public generally believes science is an arcane subject of interest only to the initiated, the support that scientific research receives from public sources requiries that it also be considered a public concern. Moreover, because scientific and technological resources comprise a significant national asset which affects a nation's economic, industrial, social and military capabilities, the management or governance of this resource should be a matter of both public and policy concern.

The management of science is a comprehensive term that can apply to a variety of activities, ranging from direct supervision of a single research project to the establishment of national policies that either directly or indirectly affect the scientific enterprise. Moving between these two extremes, management is involved in organising and maintaining academic departments, universities and research institutes. The purpose of this chapter, however, is to present an American perspective on the problem of managing science through the formation of national policies for scientific research and through deliberate planning for the development of scientific disciplines. Therefore, in this broad context, the word 'management' encompasses the idea of governance rather than supervisory direction. It assumes that governance takes place with the participation of the governed. The rationale for this position will be discussed later.

National policies have influenced science for many years. Although the nature of the ties between science and government is currently changing, the role of national policy in the conduct of science has not diminished. Science policy (including budget policy) undergirds and frequently shapes the more direct management of scientific research institutes and universities. In addition, it influences the way particular projects are conducted and it is instrumental in both developing new fields and maintaining cumulative research activities. For this reason, a broad focus on the management or governance of science at the national level is valuable. Still another reason for such a focus is that there are important issues in national science policy in both the United States and the United Kingdom – many of them related to the financial role that the national government plays in research – that must be resolved over the next several years. In this situation, better management through science policy will be needed to deal with these difficult issues.

This chapter will discuss the nature of the problems currently affecting both science and science policy: it will examine some of the larger issues that must be considered in responding to them; and finally, it will propose an approach to planning for specific fields of science in the next decade. The discussion will build on the recent experience of the United States. I take this approach in full appreciation of what are quite significant differences between nations and with no intention of arguing that the United States should serve as a prototype for science policy development in other nations or that the science policy problems or or solutions to those problems will be the same in any two countries. However, despite the evident differences, it is useful to acknowledge the experiences of other nations in developing strategies for the future. In science policy, the close ties over time between the Unites States and the United Kingdom and the similarities between the two countries make this a particularly valuable exercise for both of them.

CURRENT PROBLEMS

In both the United States and the United Kingdom, the cost of scientific research has risen sharply over the past several decades. A major reason for this increased cost is the significant changes that have occurred in the nature and scope of research. For example, in the past decade, empirical research in the social sciences has increasingly been concerned with the dynamics of social and economic processes over long periods of time. This research requires data bases that are either longitudinal in design or available in successive cross-sections over time. The former consist of surveys that revisit the same individuals or households over a number of years; the latter, the same survey conducted on successive samples of the population over a number of years. In both cases the annual cost of data collection is expensive, and since the research value of the data increases with each successive re-survey, these types of data collection require a long-term financial commitment.

The rising cost of data bases is not confined to the social sciences. Because of recent advances in molecular biology and the complexity of research questions, data bases in the biological sciences are also absorbing increasingly large sums of general research funds previously available for other research purposes. Similarly, the growing research interest in global environmental change is resulting in an explosion of data collection in the geosciences, ranging from historial land-based and marine meteorological data sets to remotely sensed data, both satellite and *in situ*. New data collection is also being planned on UV–B irradiance and its effects on plant and animal populations and on changes over time in the structure, dynamics, genetics and habitat associations of both terrestrial and marine ecosystems.

A second reason for the rise in research costs is research instrumentation. Many of the advances of recent years in biology are due to the growing technological sophistication of biological instrumentation. Moreover improvements in computation ranging from the increased power of personal computers to the use of complex parallel processing and supercomputers have led to significant improvements in research capacity across all fields of science. Yet these innovations in instrumentation, however valuable, are expensive. Despite the widespread belief that the cost of electronic equipment would decline over time, the average cost of a personal computer, like the capacity of that computer, is higher today than it was ten years ago. That the computer is far more powerful today than it was ten years ago reflects both the growing sophistication of science and the increased performance requirements for computers. It does not mean that fewer computers are needed. Moreover, despite the significant investment in instrumentation for research over the past decade, computation needs will continue to increase. In the 1990s, not only research but also education and training will require more sophisticated computers if the nation is to remain competitive economically and scientifically. This means there will be a need continually to increase the national investment in technology for research and education.

The third cause of rising research costs is the rapid obsolescence of research instrumentation and the aging of science and engineering facilities. According to a survey conducted by the National Science Foundation in 1982, US universities owned an estimated 46 000 large instrument systems for research and education. By 1985, about one-quarter of those instruments systems in the physical sciences, engineering and computer sciences were no longer being used. Moreover roughly one-half the instrument systems in use on university campuses in 1985 had been purchased since 1982.[1] Not surprisingly, in the seven years from 1980 to 1986, universities and colleges doubled their expenditures on instrumentation, with about half the total expenditure coming from current operating funds rather than capital funds. In part because of the addition of new instrumentation, research facilities are now inadequate. A survey of research administrators in the United States in 1986 found that 96 per cent did not have sufficient space to conduct research. Moreover roughly half the research administrators characterised the condition of extant research facilities as poor or fair.[2] Updating research laboratories will increase research expenditures still further.

While the need for research funding, instrumentation and facilities has been rising, national research budgets have also been growing, but at a considerably slower rate than earlier. Support for basic research by the US federal government grew by only 4 per cent per annum, in constant dollars, between 1980 and 1988. This is in contrast to the 14 per cent per annum growth rate (in constant dollars) between 1950 and 1967. Moreover federal support for research and development conducted at academic institutions between 1980 and 1988 grew even more slowly, with only a 3.7 per cent annual rate of increase over those years.[3] In some fields of science, research support levels in constant dollars were static in the 1980s. In others, they actually declined.

This situation has led to a gap between what scientists feel they need in terms of research funding and what they actually receive. It is increasingly difficult for scientists to obtain sufficient research support, no matter how excellent their ideas, and in all fields of science excellent ideas routinely go unsupported because of the shortage of research funds. The reasons for this are partly economic and partly political. Unquestionably, the United States is facing serious problems in meeting the demands of the growing federal budget. Recent legislation, particularly the Gramm–Rudman–Hollings Act which limited the annual budget deficit to a fixed amount, has curtailed federal spending significantly. Since many of the annual obligations of the federal government are not discretionary, the remaining items in the budget, including scientific research, must be reduced below the levels that either the Administration or Congress would choose, if the legislation were absent.

A second reason for what has become a shortfall in research funding is that the political environment within which science operates is changing and past experience as to how scientific research should be financially supported no longer applies. The postwar 'contract' that mediated relations between science and society in the United States for many years is breaking down.

Scientific research emerged from the Second World War as a highly-regarded activity which US political leaders believed would influence and assist the nation socially, economically and militarily. This attitude was a direct result of the successful scientific collaboration with the military in the war effort. The collaboration extended from such visible endeavours as the Manhattan Project, which produced the atomic bomb, to the less centralised but critical support with which social scientists provided the army in manpower selection, testing and training. Following the war, there was a political consensus in the United States that it was in the national interest to support advanced, fundamental scientific research at the nation's colleges and universities and, over the next several decades, a number of federal funding agencies were established to accomplish this purpose. The national research effort was intensified after 1957, when the Soviet Union put a satellite into space. The US public then began to link science with national prestige, as it had linked science with military strength for over a decade.[4] The unwritten 'contract' was in place.

With the continued expansion of the US economy in subsequent years, financial support of scientific research grew as the overall budget grew, and scientists came to expect rising research budgets almost as if regular growth were part of a natural or automatic process. Each year the US Congress approved large budgets for scientific research, with very little dissent. Even if the Members of Congress who voted to increase these research budgets could not understand the research activities for which public funds were being requested, which was often the case, they saw little reason to question the legitimacy of science's claim to the nation's financial resources. The result was a tendency within both the Congress and the Administration to provide financial support for research while making few specific demands of scientists. This widespread belief that science was good for the nation led scientists to assume that they were independent of the political controls of the society which supported them.

The situation has since changed. Although neither the existence nor the integrity of the postwar funding structure for scientific research is now being threatened, the federal budget is no longer as expansive as it was and it cannot be stretched much further. Moreover the close link between military security and fundamental scientific research no longer seems as important. Research budgets have become merely one of many hungry claimants on federal funds, and Members of Congress (as well as the Administration) are forcing scientists and their government spokesmen to defend and justify specific uses of public funds.

The mystique of science that governed the relationship between Congress and the scientific community after the Second World War is now visible only in the support given to spectacular science projects that can be tied to national prestige, such as the superconducting super collider or the space programme, and in the financial support extended to biomedical research. Fundamental scientific research is not treated so generously. A cursory look at research and development statistics in the United States indicates the priority accorded to these various areas of research. In fiscal year 1988, the federal government spent $59 104 million for research and development. Of this amount, $50 858, or roughly 86 per cent, went to national defence, health and space science and technology. Only $2160 million, or 3.7 per cent was spent on research and development in general science.[5]

General civilian science budgets are now subject to the same scrutiny, budget stringencies and political trade-offs as such other budgets as housing, veterans' benefits, and transportation. Equally important, the scientific community is viewed in Congress as not elevated above the common weal but as just another interest group on the political scene.

The declining budgets for research in the 1980s and the increased need for expensive instrumentation have given the scientific community the unpleasant feeling of being mendicants in an alien culture. Yet, before scientists spend too much energy longing for the return of the Golden Age when they were able to conduct research unhindered by external pressures or constraints, they should remember that sociologists and historians of science have found that science has always been influenced by the society within which it operates and its support, whether this is recognised at the time or not, is generally tied to the needs of either government or the private sector, or both.

To acknowledge the interdependence of science and society does not diminish or compromise the value of basic or fundamental research. Neither does it underestimate the role of intellectual curiosity as a motivating force in science. Rather, it places the corpus of scientific activities over time in the context of broader national concerns and recognises that scientific research is part of a complex of social activities. From a social perspective, a healthy scientific enterprise, including both research and training, is a highly valuable resource in modern industrial nations.

The critical factors that govern this interrelationship between science and society remain, first, that scientists have demonstrated that they are capable of responding to social needs and problems in their research, and, second, that science is not self-supporting. Modern science requires financial resources at a scale that prevents scientists from obtaining significant support for their research outside the government and large industry. Because of these close financial ties between science and external (largely governmental) sources of funding, the health of the scientific enterprise is not only dependent upon the research expertise of the scientific community, but also upon their exercise of political and management skills in obtaining support for scientific research.

A RESPONSE TO CURRENT PROBLEMS

In order to maintain constant levels of support for scientific research or even to increase that support when appropriate to take advantage of new and often costly instrumentation and computational power, scientists are told, first, that they must use existing resources more wisely and second, that they should set priorities among fields of research. At first glance, this is sensible advice and it is difficult to argue against it. Yet when scientists consider how to manage existing resources more effectively, they find that resources in most fields appear to be stretched beyond what is reasonable. Research projects in the United States are typically under-funded according to the standards of ten years ago, and the marginal gains from new economies appear to be slight because so much of the increased cost of research is a result of fixed-cost personnel, instrumentation and computation needs.

The second approach scientists are told to take to accommodate funding constraints is to set priorities among fields of research for the limited funds available. This is a difficult, if not impossible, task for researchers, because in their own careers they have already identified the research fields they think are most promising or important. There are no generally accepted external criteria that can be used to set priorities among new and emerging fields of research, nor is there a logical internal hierarchy among scientific disciplines to guide priority setting. As a consequence, setting priorities across disciplines or fields of research can be dangerously divisive within the community of science.

Most scientists would also argue that it could weaken science. The cost of maintaining the prevailing diversity across scientific disciplines is clearly greater than supporting only a few disciplines. But, because science is not predictable, this very diversity is part of the strength of science. With a broad and healthy disciplinary structure in the research community, new discoveries (such as the recent advances in superconductivity or the more disputed work on cold fusion) may emerge from unexpected disciplines or from unlikely research laboratories. Moreover there is a potential for scientific advance both through the creative interaction of various specialities and through sustained attention to more focused areas of research. Over the long run, new discoveries are less likely to occur in a scientific system characterised by heavy concentrations of research funding in one or two disciplines or lines of research at the expense of fields that are perceived as less central because their relevance to some budgetarily defined canon of intellectually, technologically or even epistemologically acceptable lines of inquiry is not immediately discernible. Science would be impoverished if, in setting priorities, scientists were forced to eliminate research activities in fields that are currently not in the forefront of scientific attention.

Research funding priorities have generally been set by those who hold the purse-strings, rather than by those who conduct the research, and this will probably continue to be the case. In the United States, this involves a diverse array of parties, including, in the public sector, the current Administration and the Congress, who determine the shape and content of the research and development budget, and government officials and scientists in research agencies who work closely with the Administration in planning R&D budgets and priorities. Too often this interaction between officials in government agencies and the Administration results in a lack of integration in science planning and an emphasis on what is affordable rather than on what is most needed.

This year, however, there has been an innovative process of priority-setting taking place to identify the research agenda and prepare future budget requests in the area of global environmental change, an effort that may serve as a model for future priority-setting in science. For the first time, federal science officials in the United States are attempting to agree upon co-ordinated research priorities and budgets across both national science agencies and research fields in a newly emerging field of research. Through the Committee on Earth Sciences (CES) of the Office of Science and Technology Policy's Federal Co-ordinating Council for Science, Engineering and Technology (FCCSET), a working group of science officials met weekly, and sometimes daily, throughout 1989 to discuss and evaluate research priorities and needs in the broad field of global environmental change. Included in the working group were representatives of the National Science Foundation (NSF), the National Aeronautics and Space Administration (NASA), the National Oceanic and Atmospheric Administration (NOAA), the US Geological Survey (USGS) and the cabinet-level agencies, Department of Interior, Department of Agriculture, Department of Energy and the Environmental Protection Agency (EPA). This group determined which federal agency should provide research support in each area of research and agreed to co-ordinate budget requests across agencies to conduct the research over a five-year period. Drafts of the working group reports and research agenda were periodically given to members of the US Committee on Global Change of the National Research Council to obtain comments and guidance from the academic research community.

The priority-setting process was difficult and lengthy, but the goal of mounting a nationally co-ordinated scientific research programme on global change and the realisation that such a research effort was by definition multidisciplinary and too great for any single agency overcame the narrower ambitions for specific fields of research or specific agency budgets. The resulting agreements were published as

Our Changing Planet: The FY 1990 Research Plan, a document released in September 1989. This research agenda will be updated annually. The success of the CES effort has led to discussions as to how this process can serve as a model for cross-disciplinary priority-setting in other fields of research.[6]

Members of the scientific community can influence research budgets and broaden research policy by becoming more directly involved in science policy discussions. Because the problem of research funding derives in part from the breakdown of the postwar consensus that scientific research strengthened the nation (and, in particular, that it strengthened national security), the scientific community should work towards the development of a new public consensus about the role of civilian science in the nation and the role of the state in promoting science. This will not be easy. Public involvement in science policy and even public advocacy on behalf of science are activities in which scientists have had little experience and which they have little inclination to attempt.

Moreover the parties at interest – the scientific community and the political/government community – differ greatly in their goals, their means of operation and the basis on which they define their actions. Unlike scientists, civil servants and policy-makers are ultimately accountable to the public. For this reason they must be persuaded repeatedly that it is in the public interest for scarce funds to be allocated to civilian science ahead of other worthy uses of those funds. This need to be persuaded of the public interest, a reasonable position, becomes more critical (and more problematic for scientists) as public funds become scarce, as they are today.

In contrast to those in academic environments, civil servants and policy-makers are held to be a working standard that emphasises conformity to externally imposed rules. Scientists believe that conformity to external regulations, outside the realm of good scientific practice, hinders creativity. They believe in the importance of untrammelled inquiry, which is frequently accompanied by untrammelled work practices. Scientists, with good reason, believe in the value of opportunistic activities: that is, the value of changing course when useful and following up the unexpected. They also believe that their activities are both superior to and more important than those of most other people and that these activities are justified as advancing knowledge rather than pursuing the public good.

The differences in orientation between those who make decisions about research budgets in government and those who depend on those research budgets for research funding do not augur well for working out a new consensus between science and society in a time of fiscal stringency. None the less it is important to attempt to do so, however difficult this may be. If science is considered a national resource, then science policy should be of equal importance to the scientist and the government official. Both can – and should – make a contribution to the development of policies to maximise scientific and technological advances. Similarly, both the scientist and the government official have a stake in seeing that science is managed so that it can make a useful contribution to the economy and to the social welfare of the population. This contribution becomes particularly important now, as we see the growing social, economic and national security importance of scientific research in non-military areas such as the environment. The task of reconciling scientific and policy interests is not impossible, for despite the differences between the operating style of the scientists and that of the civil servant, there are significant areas of common interest, backed by common goals.

An important step in moving towards a new consensus about the role of civilian research in the nation is not only to enlarge the role of working scientists in science policy, but also to ensure that discussion about the role of science in society includes the lay public. In some sense, both scientists and policy-makers have been acting as if the general public were involved in the discussion of the difficulties in conducting research under increasingly more stringent budget conditions, but the table at which the discussion was held was like the table at the Mad Hatter's Tea Party, where the guests moved from place to place until all the food was gone. In short, if scientists are to argue that scientific research is a public benefit, they must bring the public to the table.

To play a larger role in the development of national science policy, scientists should become advocates for the idea that science is a national resource that will be critical in meeting the needs of the 1990s. In the next decade, civilian research, in particular that which deals with the global environment, the aging populations of industrial nations and the social and economic needs of ethnically mixed populations, will require much greater resources. Advocacy on behalf of this research will require a reorientation in the way that scientists think about what activities are important to science. For example, scientists must learn more about public decision-making, needs and values. They must also be prepared to discuss the issues behind scientific needs and values with many audiences. Finally, scientists must recognise that science policy is not simply concerned with scientific research but extends to the full range of institutions and individuals involved in the scientific enterprise. Too often, when scientists speak about scientific research, they refer to a highly visible, elite activity in universities and academic institutes and forget that this research activity is dependent upon sustained effort in elementary and secondary schools, in industry and in the government. Because the future vitality of science depends on concerted efforts in all these areas, discussion about the role of science in the nation cannot be confined to research laboratories and higher education.

It will not be easy for scientists to become public advocates. The process requires that they add to their research and teaching responsibilities the responsibility for educating policy-makers and the public about the importance and public benefits of research and technology. However, if the scientific community is successful, it is reasonable to expect greater political consensus on the importance of science. There may also be greater appreciation for the role of science and technology in strongly improving the quality of life for the nation's citizens. This broader framework for discussion of the significance of science in a wide range of activities may well foster greater appreciation of the importance not only of applied research but of fundamental research as well.

There are many who would disagree with this strategy. Scientists are accustomed to operating on the assumption that science is an intellectual activity located in some twentieth-century chain of being far above politics, which is considered to be uncertain at best and unsavoury at worst. Many scientists would prefer to keep science in the laboratory. Unquestionably that would be easier for scientists, but science became part of the political process when scientists first began to accept public monies for research. At this point, it is inextricably part of the political system and it would be impossible for science to return to a pre-political status in the United States.

Ultimately future support for scientific research will only be obtained if science is widely recognised as being in everyone's interest. This means that the general public, as well as those in positions of responsibility in the state, must be persuaded that science is a valuable investment of public funds. Because no one can make that case better than scientists themselves, it falls to scientists to do so. It would be irresponsible to ignore or to refuse the task.

PLANNING FOR THE NEXT DECADE

Even if scientists are successful in fashioning a new consensus about the role of science in national life, we would still be faced with the fact that constrictions in science funding are real and are likely to continue. Moreover public goals will not change immediately. As a consequence, in addition to undertaking a greater role in public discussion of science policy, the scientific community must simultaneously undertake more focused internal planning for both disciplinary and interdisciplinary research. Only if this dual task is accomplished will science prove strong enough over the course of the next decade to meet the challenges that will certainly arise.

Under such circumstances, it is vital to take the perspective of what Gramsci would call the 'long period' in assessing and planning for the needs of scientific research. That is, concern for the long-term development of science should outweigh concern for the next research project. To navigate wisely through the difficulties of the 1990s, it is essential that scientists begin to study the *social science* of their fields. Therefore this perspective of the 'long period' should begin with an examination of the management, sociology, demography and political economy of a field of science; it should involve consideration of the balance between the need for research infrastructure (the collective resources of a field of science) with individual project support; and it should consider the growing need for scientific and technical personnel, the human resources of science, over the next decade.

There is an understandable tendency for scientists to know more about the content of their discipline or speciality than about its functioning. Yet ultimately the health of every field of science depends upon a conjunction of non-substantive factors, including: the numbers of trained researchers and graduate students; the quality of undergraduate departments that produce new recruits for scientific careers; the availability of instrumentation; training for both students and scientists in advanced technologies for research; and the capacity of the field to support and reward interdisciplinary and multidisciplinary research as well as traditional disciplinary research.

Because it is not currently possible to obtain new research support for all good research projects, scientists must seek ways to create a more effective environment for research at current resource levels at the same time as they continue to work to increase resource allocation for scientific research. Creating a more effective research environment is best accomplished by an improved understanding of field-specific requirements of research and the conditions under which research flourishes. For example, scientists should learn more about the statistics and functioning of both the R&D system and their own fields and the culture of graduate and undergraduate departments. They should explore new ways of managing instrumentation resources and flexibilities in sharing those resources within and across laboratories or even institutions. They should also examine the adequacy of current means of communicating research findings, such as existing publications, and the range of electronic alternatives. Scientists also need to know more about the way the research system functions in the laboratory and in the university, its weaknesses, its flexibilities and its demographic composition. Research in science policy studies can provide substantial guidance here.

Some research fields have already begun this type of assessment, but the practice must become more widespread and more routine. The need is not for special studies to determine whether a discipline is on the brink of extinction or whether it has another decade of fruitful life ahead. Rather, every field of research requires regular monitoring so that problems can be identified or anticipated before they reach the point of hindering research in the field.

A second task is to balance support for specific research projects within a field with collective research resources, that is, the technical research infrastructure that encompasses new instrumentation, data bases and other multi-user resources. In the initial stages of its development, a research field frequently requires funds for activities that permit researchers to come together to discuss rapidly-evolving research agenda in addition to training and project support; later there is a stronger need for continuing shared research infrastructure, such as data bases and major instrumentation centres. Throughout the process there must be investment in instrumentation.

Derek Price's dichotomy, the balance between support for big science and that given to little science, must be addressed in every field. This balance becomes very difficult to discuss in times of budget scarcity, when there is intense competition for research funding. There is no easy equation that can be used to determine a balance between the two types of science, and most fields, admitting differences in scale from field to field, contain elements of both big science and little science. The balance between the two will differ both across fields at any point in time and within a single field across time. What is critical, however, is that the need for balance should be recognised, both by those who do the research and by those who determine research policies, and that the decisions that are made regarding resource allocation be based on scientific criteria and potential research use by the community of scientists.

The third topic, human resources, is perhaps the single most important problem facing science today in both the United States and the United Kingdom. The health of science and the growth of a nation's capacity to respond to changing economic and environmental needs are dependent not only upon the strength of the ties between scientific research and public need, but also upon the size and the quality of the scientific and technical workforce. In the United States there are not enough students currently training to become scientists and engineers to meet the research and educational needs of the nation in the 1990s. Moreover faculty shortages are anticipated in most fields of science beginning in the late 1990s. The situation is complicated by the fact that US performance in science and maths is declining at the secondary school level, so that the pool from which new scientists and engineers will be drawn is shrinking. In order to increase the numbers and improve the quality of the next generation of scientists and engineers, the United States must improve the undergraduate and secondary school curriculum in science. In addition, there is a need for research on the incentives and disincentives attached to careers in science and technology, to understand why students are (or are not) attracted to science.

The National Science Foundation has responded to these problems by seeking to recruit more women and minorities to science. Although women and minorities have not traditionally been attracted to careers in science and engineering, they could constitute a new source of scientific and technical personnel if they can be persuaded that they have reasonable chances of advancement in these careers. If the United States succeeds in this attempt to broaden the demographic composition of the scientific community, science will be strengthened. If we fail, we will face critical shortages of personnel in ten to 20 years. Enlarging the demographic pool of scientists and engineers and providing equal opportunities for them to advance in fields that have not welcomed them in the past is one of the principal tasks facing the US scientific community in the next decade.

CONCLUSION

This chapter began with a discussion of the management or governance of science, based on the premise that the informal contract between science and society that developed from the success of the military contributions of science during the Second World War has dissolved. It ends with a discussion of the internal management of science that assumes that scientists must take an activist stance both in their relations with the broader public and in their own fields. To the extent that science remains an academic activity, this definition of the public character of the scientist's role violates the commonly accepted symbols of the academy. Yet, however unpleasant the idea seems, the 'ivory tower', which has long epitomised the distance between academia and society, is an archaic construct in science. For good or for ill, the future of science lies in the recognition of its integration in society, not its distance from broader social values and purposes. Science, I would argue, is flexible enough and the scientific community is strong enough to adapt to this change in orientation. The next step is for scientists to accept the challenge of this new set of responsibilities and opportunities.

NOTES

1. National Science Foundation (NSF) data.
2. National Science Board, *Science and Engineering Indicators – 1987* (Washington: US Government Printing Office, 1987) pp.256–8.
3. NSF data.
4. The arguments for federal support of scientific research were first made in Vannevar Bush, *Science – The Endless Frontier*, a report to President Franklin D. Roosevelt originally written in 1944 (reprinted Washington: National Science Foundation, 1980). For a discussion of the growth in federal support for scientific research during and after the war, see D.J. Kevles, *The Physicists; The History of a Scientific Community in Modern America* (New York: Vintage, 1971 and D. S. Greenberg, *The Politics of Pure Science* (New York: New American Library, 1970). W. A. McDougall, *...the Heavens and the Earth: A Political History of the Space Age* (New York: Basic Books, 1985), discusses the role military security played in securing federal support for science in the postwar period.
5. NSF data.
6. Committee on Earth Sciences, *Our Changing Planet: The FY 1990 Research Plan* (Washington, 1989).

5 The Management of Pure and Applied Science Research in Academia

ERIC ASH

INTRODUCTION

Management is concerned with optimal deployment of resources in order to attain some defined goal. I believe that many research scientists would argue that the goals of research are so diffuse that it is a subjective judgement as to whether or not one has reached them; that the whole concept of management can be applied to the research process only in a metaphorical sense. They might argue that, in a society which has seen a growing emphasis on efficiency, there are limits to be drawn. Should we attempt to manage the production of music, of poetry?

However we are not free to conclude that management has no role to play in research. First, were it so, my discussion would now be at an end, whilst in fact I have barely started. Secondly, anyone with experience of working in research laboratories in industry, government or academia will have encountered the grossest examples of mismanagement. If it can be done so very badly it is surely reasonable to suspect that it can be done only slightly badly. And, if you are an optimist, perhaps it can even be done well.

RESEARCH SEMANTICS

Scientists should define their terms, but nothing is more boring than when they set about doing it. If, however, we are to cast any light at all on the management of research, we cannot wholly escape engaging in some dissection of that overworked word, 'research'. It is used to describe the efforts to understand the ultimate realities of nature and, almost in the same breath, to illuminate the preferences of a population for a consumer product or a political policy.

There has been much discussion in recent times on a taxonomy of the research process. One approach recognises a sequence starting with 'pure' research, motivated primarily by curiosity, by the inherent and possibly even irrational desire for knowledge and understanding; 'strategic' research which, whilst not directed at the emergence of utility, is in a field where there is a reasonable expectation of economically useful applications; 'applied' research, in which known basic science is brought to bear on exploring the possibility of satisfying new perceived needs; 'developmental' research, where the problem to be solved is clear, and the outline of the solution evident, but where there remains a gap, sometimes very wide, of 'reduction to practice'.

There is, I believe, general agreement that some such classification is meaningful even if less than perfectly sharp. There is also a view – I would describe it as a myth – that there exists a unidirectional flow of ideas from one category to the next. On this model, progress starts with curiosity-based, pure research, with an understanding of a new scientific principle. It is followed by the recognition that the principle could be useful, and further strategic research then identifies a specific area of application. Applied research, followed by a developmental stage then leads to a desirable product. This sequence is indeed evident in the history of science and technology. Faraday's discovery in 1831 of electromagnetic induction, followed by the demonstration, very shortly thereafter, of the dynamo, which half a century later led to the widespread use of electric lighting and electric motors – this is a classic example of a sequence of this kind.

Less recognised by the public at large and, just possibly by some government ministers, is that there exists a reverse flow along this chain, which is at least equally important. The idea is embodied in the perception that 'science owes more to the steam engine than the steam engine owes to science'. Steam engines were invented and brought into use before the basic laws of thermodynamics had been discovered. But the fact that steam engines were around, and most evidently worked, led to some key questions – such as what is the maximum efficiency of a steam engine – questions which eventually led to the emergence of a new branch of science, thermodynamics.

A more recent example is to be found in the discovery of holography. In the early 1940s, Denis Gabor was concerned with the improvement of the electron microscope. Now the basic problem with electron microscopes, then as now, is that the lenses which one uses to magnify images are, in a fundamental way, vastly inferior to the lenses which we can devise for optical instruments. Electron microscopes still win out because the wavelength associated with the electrons is so very much smaller than the wavelength of light. They win out, but not by as large a factor as one would achieve given better electron lenses. Gabor was enmeshed in this 'developmental research' activity. However it led him to a very simple question, the ability to ask such a question being one of the hallmarks of genius; if all we have are poor lenses, cannot we find a way to image without them? That question led him to the discovery of holography, which has proved to be a field which offers a great variety of important applications (though, oddly, probably not to electron microscopy). However, in addition to its utility, holography is recognised as a fundamental concept of great theoretical importance and of inherent beauty. It must be seen as a fruit of 'pure' research.

The recognition of this 'reverse flow' of ideas is, I believe, important in devising strategies for the effective management of research – anywhere, but not least in academia.

GOOD OR BAD RESEARCH

There is one further classification which I would like to touch on which I believe is important in devising a strategy for the management of research. Should we undertake good or bad research? I would like to suggest that the answer is not quite as obvious as it might appear at first sight.

Harold Clurman, who for some 50 years had such a dominant influence on theatre in the United States, once said that 'the history of theatre is the history of lousy plays'. He was not of course seeking to encourage anyone deliberately to write a lousy play; his message was simply that there has to be a great deal of theatre, a great number of writers for the theatre – most of whom will be of modest talent and achievement – if we are to have any hope of witnessing the flourishing of exceptional talent.

The analogy to scientific research is evident. Yet universities have been asked by government, 'Are you sure that the least important of the researches carried out within your portals is viable? Would it really matter so very much if you were to cut the bottom 10 per cent?' The answer is of course quite clear: It would matter very little. The difficulty lies in identifying that 'bottom 10 per cent', and to do so before the resources on which it relies have been allocated. True creativity in the sciences is sometimes as hard to gauge, at the instant of creation, as it is in the arts. Beethoven's late string quartets were seen, by contemporary critics, as an indication of his failing powers.

I do not wish to exaggerate the point. Peer review actually works not badly. But it is an approach which can, at best, claim a reasonable *statistical* success. It will, however, inevitably lead on occasions to research which is not distinguished; less often, it will discourage truly innovative ideas because they are ill-expressed or ill-understood.

When one has done all that one can to select on various criteria, one will at times witness truly exhilarating results. One will also see a lot of lousy plays. You cannot have the former without the latter.

ACADEMIC RESEARCH MISSION

I hope that I can regard it as common ground that research is part of the essence of the concept of a university. Happily some suggestions a couple of years ago, that one might envisage a university devoted uniquely to teaching, have faded. All universities have a research involvement, though the emphasis on this aspect of our total mission will vary from one institution to another. We are concerned with all the categories of research that we have already listed; we seek to discover the new; to apply what is known for economic and social benefit; and, very importantly, we are concerned with teaching (I dislike the word 'training' in this context) young people who may want to spend a part of their career in research.

There has been a great deal of debate, some of it well-informed, on the relative emphasis which we should give to these different aspects of the research mission. I have myself already said all I know about the matter – or possibly a shade more than that – on a previous occasion.[1] Our task today is to focus on management rather than purpose – though of course these two concepts are intertwined. I will not therefore dwell further on relative research priorities, but rather make some comments on what must be the first consideration in management – the resources which are to be brought to bear on the task.

RESOURCES: FUNDING

What can be achieved in university research depends of course on the level of funding. I do not wish to say very much here on the magnitude of the government support for universities. Were I to do so, I would tell you that it is too small; that the evidence that our competitors in Europe and elsewhere are spending proportionally more is persuasive. This is part of a continuing debate which will, and in my view should, continue. A succinct discussion of the present situation has recently been presented by Sir David Phillips, Chairman of ABRC.[2]

Overheads

Research costs can be split into direct costs, for paying salaries, equipment and materials, and indirect costs, which might include heating, lighting, cleaning maintenance, accountancy, purchasing and – management. The indirect costs are normally defrayed by means of an 'overhead' on the direct costs; nothing very mysterious here, particularly in industry. Nor is it particulary difficult to calculate on an overall basis what the correct overhead percentage amounts to, for a particular organisation. So, for Imperial College, it is expressed as 120 per cent of salary costs, which translates into something like 75 per cent of all direct costs. That figure emerges from calculations conducted by our accountants. The figure for other universities is similar – it tends to run between 100 and 150 per cent. Nor is it very different in the United States. So, for example, at MIT it works out at 100 per cent of all direct costs, corresponding to the upper limit of the figures for UK universities.

Oddly, in fact very oddly, this concept has proved to be one of great difficulty for some of our industrial firms, notably some of the largest. There is some sort of feeling that Corporation taxes have already paid for the universities; that, whilst the direct costs are of course acceptable, the infrastructure should be free. Now a back-of-the-envelope calculation would show that industry has paid for only a few per cent – certainly less than 10 per cent, of the university infrastructure. It should not be difficult to agree that not only research workers in industry but also those in universities need heat and light. I have at times felt like asking 'If we are pricked, do we not bleed?' But facts are not necessarily victorious when ranged against belief.

I am glad to say that there is a very encouraging trend towards understanding and acceptance of this problem; an increasing number of our industrial colleagues have grasped the situation. I might add that it is also very encouraging to note that our collaborators abroad – in the United States, in Japan and in Europe – also find the position we have adopted in academia as natural and equitable.

There is, however, a long way to go. Recently the measure of overhead recovery achieved by different universities was published in the form of a ranking table. Even the best performer, Salford University, recovered, on average, less than half the true overhead cost. I cannot remember off-hand how the ranking sequence went, and anyway not everyone likes these ranking results. I do, however, just happen to remember who came second – a place called Imperial College.

So the industry scene is improving. I wish I could say as much for government departments. They are most disinclined to pay more than half the overhead costs. Now what does this actually mean? We obtain a substantial part of our total income from the UFC. The grant (soon to be a contract) is there to support teaching and non-contract research. If we recover inadequate overheads for contract research from, say, the Ministry of Defence, what it really implies is a subsidy which the Ministry of Defence has *forcibly* extracted from the Department of Education and Science. Has the Treasury grasped this? In their demonology this must surely be regarded as an unforgivable evil. So why is it tolerated?

The government has at times suggested to industry that they might do more to support at least segments of the university system. Encouraging government departments to recognise the actual cost of research contracts which they place would be an inspiring example.

Debts

All organisations have debtors, and so do we. With luck, in a powerful corporation the debtor account is balanced by the creditors'. It is one of the hazards of small start-up ventures that this balance is upset, to the disadvantage of the organisation. Universities tend to experience the world as does a relatively small company. The debts tend to prevail. What is particularly painful is that, the greater one's success, the greater this problem. In the case of the research councils it tends to be built in by the procedures which essentially require, first, the incurrence of the expense, then the submission, then the controls and finally the payment. Yet much of the expenditure is totally predictable – for example, that devoted to salaries. I believe that it would be possible to devise procedures which, while protecting the public purse against any abuse, would put less strain on the university cash flow.

Problems with government departments are very similar. On the whole, in our experience, the recommendation that to give is better than to receive, is not one which has gained much credence in government departments. An extreme example hit us last October. A Ministry – which shall be nameless – suddenly refused to pay any of our bills, without giving any reason for their reluctance to do so. On enquiry it turned out that the Ministry insisted that it had made a contract with Imperial College of Science and Technology. Yet, the bills came from some unknown place called Imperial College of Science Technology and Medicine. We confessed that we had indeed merged with St Mary's Hospital Medical School. You will be glad to learn that payments have recommenced.

Intellectual Property Rights

Some academic research is unlikely ever to have a commercial value. Astronomers are more likely to find manna in the heavens than on earth. High-energy particle physicists seem rather remote from economic utility, though there can be useful spin-offs. It turns out, for example, that some of the work at Imperial College on particle detectors is proving useful in CAT scanners. Even pure mathematics is not quite safe from utility, as was noted by Lord Flowers in a speech in the House of Lords:

> 'Pure' is not synonymous with 'useless' as some seem to think ... For example, secure message encoding depends on the properties of very large prime numbers, a subject fondly thought by mathematicians, not so very long ago, to be utterly useless. The commercial value of this discovery would surely cover the cost of all the

mathematical research since Euclid initiated the theory of numbers.

However, more usually, commercially valuable discoveries emerge from strategic, from applied and from developmental work. Who owns the intellectual property rights (IPR) of a discovery? It has to be said that, until a few years ago, there was rather limited interest in this topic, at least in UK universities. It is true that occasionally there were major inventions which were so perceived by the inventor at the time, which were patented and exploited. However, on the whole, academics carry out research because they love to do so. They also want to publish – to communicate with other groups and, of course, in support of career advancement. Academics really only get fussed about money when they have been squeezed for one and half decades.

At the same time, the university institutions did not put any great emphasis on securing IPR or on how it should be shared between the organisation providing the funding, the university institution and the begetter of the IPR.

All this has changed. The government segment of university funding is steadily decreasing. Universities have to earn an increasing part of their living in a more direct manner. It is important to appreciate that this is necessary not so much to support applied research but to protect the pure. In short, universities have realised it to be essential to take IPR no less seriously than any high technology company. This transformation is not complete; it involves a difficult management problem – communicating in meaningful terms with all of the academic staff in a university; discussing procedures; deciding what should be an equitable distribution of the fruits of commercialising IPR. The transformation is under way and in many institutions is well advanced.

That is the internal problem. We must, however, also consider our relationship to our collaborators outside the university. Here, thanks to a government initiative of some five years ago, the situation with regard to Research Council grants has been totally and generously clarified. The grant holder and his university are free to protect IPR as they think best; to commercialise as they think best.

There are problems when we engage in a collaborative research with an industrial company. If the company pays the whole cost of the research (direct and overhead costs) it is natural that the company should own the IPR. Suppose, however, that the costs are shared

50:50. One does not have to be imbued with the wisdom of Solomon to suggest that, in that case, it would be reasonable for the IPR to be shared in the same proportion. I am happy to say that many companies understand that proposition. I am unhappy to tell you that some do not. The arguments which are advanced against what would seem to be the obvious approach are varied and exotic. Let me give you two samples. The first is to suggest that universities are innocent of the difficulties and dangers of the real, the commercial world. They need the protection of a corporation that has experience in these matters. Now that might have been fair comment a decade ago. It simply is not true now. Most universities have started companies whose sole *raison d'être* is to identify and exploit IPR. Their technical expertise will rival that of most organisations; they have a number of different avenues open to them for obtaining patent protection; they will normally have several lines of finance at the ready; they will often have overseas offices. The universities can cope.

The second argument is that, if the IPR is shared in some way, say in the sense that either party can give non-exclusive licence to a third party, then the company may find itself having shared its IPR with that of a competitor. That is of course perfectly true. However the company will have had a head start of several years before that competitor gets a smell of the invention. And if there is any lesson to be learned in the exploitation of high technology it is surely that only a dynamic advantage is much use. If the company feels that the IPR is so critical to its own mission that it wants to avoid dissemination to others, it is always open to it to buy the half of the IPR that it does not own from the university.

A number of universities are beginning to obtain significant benefits from the commercialisation of IPR. We still have some way to go before we can equal the performance of American universities which academically are on the same level – but we are rolling. The sociological transformation which is needed both in academia and in industry is happening, but there is still some considerable distance to cover.

Collaborative Research

A key feature in the manner in which research is conducted in academia and in industry has been the growth of collaborative research as originally devised in the Alvey Information Technology scheme. The key concept, as I am sure the reader will be aware, is that of several industrial companies and several universities engaging in 'pre-competitive' research. There is, as might be imagined, a spectrum of views on just how successful this scheme was. It is fiendishly difficult to assess. My own view is rather enthusiastic. I believe it achieved exploitable ideas and products in information technology. In addition it taught the whole of the research community not only how universities and industry can best collaborate, but also how two industrial firms can work together – without either stealing the silver spoons from the other.

It was a pioneering venture. It had blemishes, from which we have learned. There have been a whole raft of subsequent experiments on the same line – most of them European Community-wide. The most important national collaborative scheme currently running is the LINK scheme. Its vision differs a little from that adopted by Alvey. It is intended to have a more direct and more rapid impact on productive industry, though avoiding too 'near-market' research, which, it is rightly felt, should be the province of industry on its own.

I have been highlighting problems (but, I insist, the curable variety). Where is the problem here? It lies in the insistence by the Department of Trade and Industry (DTI) that not more than 50 per cent of the total funding should come from the public purse. Now universities have no money of their own, so that to participate they need 100 per cent funding. In a collaboration between, say, two companies and two universities, since the universities get 100 per cent and the total can only be 50 per cent, it follows that the companies get less than 50 per cent. Nor does it take a firm of management consultants to explain to us that, the greater the university share, the smaller the percentage of the research costs recovered by the companies from government. Nor would it take the same firm of consultants to explain that, if this is the deal, the companies will prefer token contributions from and to the universities.

The LINK scheme is, I believe, an imaginative and potent mechanism for enhancing exploitable research in the United Kingdom. Yet it is finding it hard to spend the money which has been allocated. The answer is simple – retain the rule that industrial companies derive a maximum of 50 per cent from the DTI; that universities obtain 100 per cent. I do not think that this seriously dents any principle promulgated by the DTI. Could not this change now be made? I note that we have recently had a change in the Secretary of State. Now I do not really understand politics but I have noticed that a change of policy does sometimes follow when a new Secretary of State takes office.

RESOURCES: PEOPLE

Which matters more, people or funding? The answer cries out for a cliché. However I believe that the question is wrongly posed. The ability for a university to be productive in research is surely the *product* of its intellectual excellence and funding. You can test that theory at its extreme: no funding, and you cannot recruit that prime number theorist, even though he needs no more than pencil and paper by way of equipment. Assemble a scintillating scientific group in any experimental discipline and they can do little without massive funding. On the other hand – and there are examples, to be found, particularly in the United States – obtain vast funding for research and couple it to mediocre scientists, and the result is a lemon, or at best several lemons. The contrast between these two extremes is that a brilliant faculty with no cash can probably find ways of acquiring it. A vast fund of money cannot turn a mediocrity into high talent. So perhaps the cliché answer – it is people that matter most – is after all appropriate.

How should academic scientists be managed? My personal stance would be to say 'barely at all'. People embark on an academic career for a number of reasons. They will enjoy teaching; they will be fascinated by research; they will probably be workaholics; they will not be particularly concerned about getting rich. There is, however, an additional trait which I believe is a factor for many recruits to academia: they want to enjoy a great deal of personal freedom. It is one of the glories of the concept of a university, that no one really bosses anyone else. When you ask just what sort of *orders* a professor, a head of department or a rector can give it turns out to be few or none. The whole system is a persuasion business.

I am not, of course, suggesting that in industry, or in government establishments management is effected by barking orders to one's underlings. *All* successful human interaction must have a basis in persuasion. There is yet a difference in kind in an organisation which has to run almost entirely on consensus from those where there is an ultimate discipline of instruction which can be imposed.

The freedom is one of the glories of academia. It is also an impediment to effecting a fast response to changes in external circumstances. Adopting a total *laissez-faire* policy is a recipe for inefficiency, if not disaster. How can one then reconcile the maintenance of the highest possible degree of personal freedom with the need to husband and to manage resources?

I was much cheered a little while ago to hear that Sir John Harvey-Jones had decided that universities were unmanageable. I had suspected something of the sort, but this endorsement suggested that the perception was more widely held and more valid than I had realised. Of course it is an aphorism and presumably, if expanded, it would have been seen to mean not as tightly, not as efficiently as ICI.

My personal conclusion is that universities work best if the centre produces a framework within which resources can be discussed and allocated and delegates authority to departments which have the widest possible freedom within the overall resources available. I have seen such a liberal regime work well, both as an academic near the coal face, and in looking at the scene from the centre. I believe further that the most successful departments in turn delegate authority down to research groups; that research group leaders transfer responsibility down to individual project investigators.

I have myself gone one stage further, in asking a new PhD student, working on a funded research project, to regard that project as a cost centre for which she or he is reponsible. It has the minor advantage of relieving the supervisor from a good deal of detailed management and the major advantage of giving a good deal of managerial education in parallel with the main purpose of the exercise – to embark on exciting and fruitful research. I do not, incidentally, believe that there is a wide enough appreciation of the excellent grounding in management that a period spent on doctoral research can provide. A student will discover very early on that, on the whole, things in this wicked world do not work; that one has to think rather hard, and sometimes talk rather fast, even partially to overcome this unhappy defect in our universe. That surely is worth some course entitled 'Management Basics'.

Before one can delegate responsibility one has to distribute resources. That to my mind is the key problem of the system of distributed management and responsibility. One will do this by echoing to a considerable extent the criteria used by the UFC in funding universities. One must, of course, be in a position of calculating the calculable loads, and this requires efficient financial and statistical management information. One must also try to build in incentives for aspiring to excellence. One will, of course, never succeed. One is trying to measure justice with an accuracy which it is simply not possible to achieve. But one can get close.

Having done this, one does avoid questions which, in the short term, always seem to me to be meaningless such as which is more important – physics research or chemistry research? Yet in the longer run a loose framework can accommodate and react to the changes which take place in science – for example, the rapid increase in the importance of the life sciences.

FINAL COMMENTS

I have at best touched upon some of the issues which concern those who have a responsibility for the management of scientific research in academia. I have concentrated on a small number of issues which concern us, chosen on the criterion that they are candidates for change and a change which does not depend primarily on increased resources. A few more of those would, of course, also be welcome.

The business in which we are engaged is complex, far more complex, I would argue, than is encountered even by large corporations. We are here to teach, to research, to support industry and commerce. We are also here because universities are of inherent importance, not only to enhance prosperity, but also to give meaning to a society that does not live on bread alone.

NOTES

1. Eric Ash, 'Higher education, industry and Government – new rules for a ménage à trois', *IEE Proceedings*, vol 135, pt A, no. 1, January 1988.
2. Sir David Phillips, 'Funding the UK science base: modes of support', *Scientific Public Affairs,* vol.4 (1989) pp.59–63.

6 Finance Policy and High Politics in a European Scientific Laboratory: The Conflicts over Financing CERN in the Late 1950s and Early 1960s

JOHN KRIGE

PRELIMINARY NOTE

CERN (European Organisation for Nuclear Research), based just outside Geneva, is a laboratory essentially devoted to doing basic research in high-energy physics. Established in 1954, it now has a staff of about 3400 people, and its facilities are used by several thousand outside visitors, not only from Europe, but also from the United States, Japan, the Soviet Union, China and the third world.

CERN's central funding is at present provided by 14 European governments – Finland plans to become the fifteenth – who share the burden proportionally to their gross national products (GNPs). The laboratory's annual budget in 1988 was some 810 million Swiss frances (about £325 million).

Britian was one of the founder members of the organisation and, along with France, Italy and the Federal Republic of Germany, has always been a major contributor to the budget. Between 1957 and

1961, her share hovered around the maximum 25 per cent permitted by the organisation's Convention. By 1988, it had dropped to about 17 per cent, corresponding to a contribution of almost £56 million.

The day-to-day management of the laboratory is in the hands of scientists who are presided over by a Director-General, himself a prestigious physicist. The laboratory's budget is drawn up by the CERN management and negotiated in the Finance Committee, which is composed of at least one science administrator from each member state. The recommendations of the Finance Committee are passed to the supreme governing body, the CERN Council, in which each member state is represented by two delegates, one a high-level science administrator or diplomat, the other an eminent scientist. Britain's representatives on the CERN Council in 1957 were Sir Ben Lockspeiser, the Secretary of the Department of Scientific and Industrial Research (DSIR), and Sir John Cockcroft, the Director of the Atomic Energy Research Establishment at Harwell. Today her delegates are Professor E.W. Mitchell, the President of the Science and Engineering Research Council, and Professor Don Perkins of Oxford University.

INTRODUCTION

Britain has always been a reluctant and hesitant partner in CERN. In 1952 she was the last of the major European countries to join the provisional organisation set up to plan the future laboratory – though she was the first to ratify the Convention establishing the organisation in 1954. Between 1957 and 1961, she stood almost alone in insisting that there should be strict limits imposed on CERN's expenditure, and it took a number of extreme measures to get her to accept a policy of growth for the laboratory. In 1968, Her Majesty's Government decided not to participate in the 300 GeV accelerator project which was then maturing – only to change its mind later and to join the venture officially in February 1971. Finally, in 1985, the British government let it be known that it was considering withdrawing from CERN, the threat only being lifted after measures had been formulated for 'improving efficiency within CERN'.

In this chapter I want to focus on just one of these developments: the debates over finance policy which occurred in 1957 and in 1961.[1] My aims are essentially twofold. Firstly, to present the results of a piece of historial research, to describe in some detail and to analyse in some depth the process of decision around the level of the CERN budget; secondly, to explore the considerations shaping Britain's attitude to CERN expenditure, to enrich our understanding of the finance policies defended by her delegates, and to show how and why those policies were accommodated to the political realities of the day. In short, I want to bring out the way policy and high politics are interwoven in an international laboratory like CERN, with particular reference to British 'victories' and 'defeats'.

FROM *LAISSEZ-FAIRE* TO EXTERNAL CONTROL: BRITAIN'S CEILING POLICY

Our story begins in 1957. In spring that year the UK delegate to the CERN Finance Committee (FC), Harold Verry, suggested that the time had come to impose some form of external financial control over CERN's expenditure.[2] Until then the policy in the Committee *vis-à-vis* the CERN budgets had been one of *laissez-faire*: the management estimated what they thought the laboratory needed the following year, and the Finance Committee delegates more or less accepted their figures and did what they could to raise the money at home. As a result, said Verry, the Committee risked degenerating into little more than a rubber-stamping body. CERN, said the British delegate, had to realise that funds for the laboratory were limited and that, just like any national laboratory, it had to set priorities, adjusting its scientific programme to the resources made available for it by the member states.

The concrete suggestion which emerged from this initiative in the summer of 1957 was that a *ceiling* be imposed on CERN's expenditure. What this meant in practice was that the administration was expected to make a two or three-year forward estimate of its needs, that the *total* allocation for that period was then negotiated with the Finance Committee, and that the final figure subsequently voted in the Council was to be fixed and binding on both parties, the laboratory and the member states.

Why did Verry make this proposal at this time? To begin with, because his delegation felt that the existing framework for allocating resources to CERN was rapidly 'becoming a complete fallacy for financial safeguard'. When the laboratory's Convention had been drawn up in 1952, it had been assumed that the major budget item would be for the construction of its two main accelerators. However in 1956/7 it was becoming clear that the costs of research, the cost of actually exploiting an accelerator as an experimental tool once it was built, were going to be far greater than was previously thought. There was a revolution in detection techniques under way which was to transform the face of high-energy physics (the development of the hydrogen bubble chamber) and at this stage no one could make any meaningful estimate of how much was needed to equip CERN. As Verry put it, CERN was working in a field where there were 'limitless possibilities in the way of research opening out almost every day ...' and in which there was 'very, very hot competition with two countries who [could] spend almost unlimited amounts, the United States and Russia'. That granted, he felt that a more rigorous form of 'external' financial control was needed over CERN's budget.

We cannot leave it at that, however; other considerations specific to the situation in Britain at the time also played a role. To appreciate their importance let us simply note that, while his colleagues on the Finance Committee shared Verry's concern about the unknown, and rising, costs of research, they drew opposite conclusions from it. Indeed they felt that, in the light of the changes taking place in the state of the art, it would be fatal to CERN's development to impose medium-term constraints on its expenditure. The healthy growth of the laboratory, they believed, demanded that the management be free to react to new developments in technique without paying undue attention to costs. Why did the British delegation think differently?

Firstly, there was the government's continuing lukewarmness *vis-à-vis* international organisations. In 1956, Verry said, the United Kingdom had contributed 'some six million pounds sterling, very largely in hard currency [to such bodies], and CERN was among the half dozen largest contributions'. It must not be forgotten, he pointed out, 'that some Governments now instinctively associate the thought of International Organizations with every-increasing budgets', and that if CERN's needs were 'to continue to command the sympathy of, at any rate the U.K. Treasury, the Organization must show that the size of the annual budget is under effective control, and based on an intelligent and realistic policy'.

Then there were the criticisms being voiced by the domestic physics community: that CERN was 'in fact l'enfant gâté' of their government. Among 'national workers', said Verry, CERN 'has the reputation of being able to get anything it asks for, as compared with the difficulties encountered by the national institutes in getting State grants'. As a result, scientists in Britain, 'animated perhaps by quite understandable envy ...' – the people in universities, Verry said, had to deal 'with bits of string and nails' and 'to improvise apparatus' – were suggesting that CERN was 'lavishly equipped'. Matters were not helped by the attitudes of the CERN staff who tended, said Verry, 'to look down somewhat condescendingly on the less well-equipped and perhaps make-shift apparatus of national institutes'.

Then, though this was not mentioned aloud by Verry at the FC meeting, there was the financial squeeze engendered by Britian's determination to develop its national programme in parallel with CERN. On 14 February 1957, the foundation of NIRNS – the National Institute for Research in Nuclear Science – was announced in the House of Commons. According to one of the key planning documents, the centre was to cost some £10m (about SF120m), £8m of which was for 'two large machines of different types', one being the 7 GeV high-intensity proton synchrotron, Nimrod.

These arguments point to a fundamental difference between the financing of CERN in Britain and that in at least some of the other member states, a difference which persists to this day and which continues to differentiate the United Kingdom from some of its partners: the source of money for CERN in the national budget. In many continental countries, membership of CERN was seen by high state officials primarily as a political gesture – in France and Italy, for example, it was sometimes specifically associated with the thrust towards European unity – and resources for the laboratory were made available by the Departments of Foreign Affairs.[3] Money for CERN and money for the domestic science effort thus came from different pots. This was not and still is not the case in Britain where there was less enthusiasm for European political unity, and the main motive for joining CERN was said to be scientific and economic: it was the cheapest way of doing physics at 25 GeV. As a result, the funds for the laboratory were initially part of the DSIR's vote; today it is the SERC which foots the bill. In practice this means that universities and national research institutes, for example, have to compete for funds from a pool which is 'automatically' depleted by the contribution made to CERN. That contribution today is over £55 million, or almost one-sixth of the total resources allocated to the SERC for 1987/8.[4] The problems raised were neatly explained by Verry over 30 years ago: 'Whether it realizes it or not' said the British delegate, CERN 'stands in a very, very privileged position when one considers that in everyone's own country scientific societies or organizations come to the Governments for money, and that they are usually either turned down or given half of what they ask for. CERN is in the happy position of nearly always getting what it asks for.'

As we have already mentioned, Verry's suggestions were coolly received by many of his colleagues in the Finance Committee. In their view this was a singularly inopportune moment to be abolishing the *laisser-faire* policy and to be imposing limits on CERN's expenditure. Given the difficulty of predicting the costs of research when new, complex experimental techniques were under development, it seemed to them that it was premature to try to impose ceilings and that the policy, if enforced, would inevitably hamper the proper development of the laboratory. Correlatively, the delegates were afraid that if ceilings *were* imposed, only to be broken, their credibility before their domestic financial authorities would be seriously jeopardised, and their task of raising money for CERN would be that much more difficult.

Despite this opposition the CERN Council imposed a ceiling on CERN's budget for the first time in December 1957. The majority went along with the idea, albeit somewhat reluctantly, primarily for two reasons. Firstly, the policy promised to solve a *procedural* problem. At the end of 1957, the CERN management put forward a budget request for 1958 that was some 50 per cent above the estimate provided the year before. Most domestic financial authorities, however, had long since set aside their CERN contributions for 1958, and this last-minute request for an increased allocation created considerable bureaucratic difficulties. The ceiling policy, by fixing a two or three-year envelope on expenditure would, in principle, stop this kind of thing from happening again.

The second, and far more important, reason why the most influential delegates in the Council 'accepted' the ceiling policy was that it was the only way to get the British, in particular, to support the budget figures favoured by the majority. That majority generally backed something like the figures being put forward by the CERN management; the British, usually along with the Scandinavians in this period, systematically advocated lower figures. Unanimity, or near unanimity, in the budget vote was achieved by striking a deal: Britain would go along with the majority figure if, but only if, it was coupled with the imposition of a two- (or later three-) year ceiling on CERN's expenditure.

One may ask why there was this desire for consensus in the CERN Council, all the more so since, *formally* speaking, there was no need for it. The budget at CERN could be adopted by simple majority, and if that majority had simply stuck to the letter of the law they could have voted in favour of their preferred figures *without* having the ceiling policy 'imposed' on them. Whence, then, their willingness to accommodate the British? Superficially, because there was a 'gentleman's agreement' among the Council members not to outvote a major contributor to the budget when settling CERN's annual level of expenditure. Fundamentally, because the leading figures in the Council, men like Bannier (Netherlands), de Rose (France), and Willems (Belgium), were convinced that unity was strength, that their and the Council's ability to foster the development of CERN depended crucially on their presenting a united front to member states' governments whenever possible. In Bannier's memorable phrase, Finance Committee and Council delegates were not simply 'representatives of the Member States ... *they were at the same time CERN's representatives with their respective governments*' (my emphasis) and the more cohesive they were the easier it would be for them 'to present CERN's point of view' at home.[5] For a man like Bannier, then, far more was at stake in the budget vote than mere money: a split vote would divide the council, undermine its power, restrict its autonomy and call into question the very future of CERN itself. It was not a step one took lightly.

THE BUDGET CRISIS AT THE END OF 1961

The ceiling imposed on expenditure for 1958 and 1959 was not kept to: set at SF100m at the end of 1957, the Council effectively agreed at the end of 1958 to boost expenditure for those two years to SF110m. This concession was granted on condition that a new ceiling be imposed for 1959 and 1960 together. After lengthy negotiations this was set at SF120m – just below the SF124m that the administration said it needed. At the same time, the British insisted that CERN's expenditure for 1960, 1961 and 1962 remain fixed at SF65m per annum.

This attempt to stabilise the budget by imposing a fixed, three-year ceiling also failed. CERN only kept its estimate for 1960 within the SF65m limit by absorbing all the reserves set aside for the entire three-year period. Its budget request for 1961 actually exceeded the

SF65m ceiling by some SF4–5m, but the Council found various means of meeting the management's request without triggering a crisis. However it was clear to all that the crunch would come in 1961. And indeed, in spring that year, the management warned the member states that they would be looking for something like SF75–78m for 1962 – way above the SF65m ceiling imposed by the council two years earlier. It was a request that was to lead to a head-on conflict between sections of the British government and some leading figures in the CERN Council, a conflict in which 'high politics' was brought into play, and from which the United Kingdom emerged bruised, beaten and forced to reappraise its policy on CERN

At the heart of the conflict lay a difference of policy between the British Treasury, on the one hand, and the French delegates to the CERN council (backed by the majority), on the other. As far as the Treasury was concerned the costs of CERN, which had been rising steadily since 1959, should now *stabilise* at least for a period of three years. As far as most CERN Council delegates were concerned, experience in national laboratories showed that stabilisation meant stagnation: to have a 'healthy scientific life' (to quote French Council delegate Francis Perrin) CERN needed to expand and its budget needed to *grow* each year with respect to the year before. It was this contrast between stability and growth that put the British at odds with the majority at CERN in 1961, and which led to an explosive debate on the budget for 1962 and the estimates for 1963 and 1964.

The detonator for that explosion was laid in November 1961 by the Treasury and the Foreign Office. On the one hand, they let is be known that they were not prepared to give CERN more than SF240m for 1962 to 1964 inclusive, as against the SF260m that the CERN administration insisted that it must have. At the same time, in an attempt to put teeth into their figures, the Foreign Office officially approached the other member states and suggested that such an overall figure, rather than resulting from a vote in the CERN Council, *be negotiated between the governments themselves* and then presented to the Council and the laboratory administration as a *fait accompli*. 'Member governments', an *aide-mémoire* from the Foreign Office said, 'should make up their minds' how much they were prepared to give CERN for the next three years and 'the Organization should realise that this total must be adhered to'.[6]

The British initiative was interpreted as a fundamental attack on the laboratory by eminent members of the CERN Council when they met to settle the budget for 1962. Dutch delegate Bannier was furious at the diminution of the Council's powers implied by the Foreign Office's *aide-memoire.* To bypass the Council by settling a budget envelope between governments, he said, would not simply diminish its authority; it would 'demolish ... the very foundations of CERN' itself.[7] French Council President and diplomat François de Rose was equally incensed by the slow rate of growth implied by Britain's suggestion that there be a ceiling of SF240m on CERN's expenditure for 1962 to 1964. This figure, he claimed corresponded to a growth rate of 2.5 per cent per annum, and was to be compared with annual rates of expansion in big American laboratories of between 15 per cent and 18 per cent, and of something like 12 per cent in France.[8] If CERN 'were to fulfil in Europe the duties which it had been given with regard to national laboratories' and to 'keep up in the competition with big laboratories in the United States and the Soviet Union', de Rose went on, 'the only suitable rate of expansion would be one comparable to national laboratories'.[9]

De Rose was not prepared to leave matters at that, however: indeed he threatened to take two retaliatory steps if Britain insisted on pursuing her avowed policies. Firstly, he threatened to advise President de Gaulle that France should not join in two new European organisations whose formation was then being negotiated: ESRO (space research) and ELDO (launcher development). And lest anyone think that his threat was idle, de Rose remarked that he had already persuaded a reluctant de Gaulle to accept that ELDO use Britain's 'Blue Streak' rocket as a launcher. Secondly, de Rose threatened to advise his government to break off negotiations with the Swiss for an extension of CERN's site into France. These arrangements were naturally 'based on the hypothesis that CERN would develop'; if it did not, if the Council voted a three-year ceiling of SF240m the next day, 'I could not go back to my government and say I continue this proposition,' said the Council President.[10]

Faced with this onslaught, and the widespread support for it in the Council, The British delegates had no choice but to back down. The Foreign Office's suggestion that a budgetary envelope be agreed between governments was not taken up again after Bannier's attack, and de Rose's threats put an end to any further talk of a three-year ceiling of SF240m. When it came to the vote, the budget for 1962 was fixed at the majority's preferred figure of SF78m, only Britain and Sweden voting against, and even then the latter implied that they might have voted with the majority if they had realised how great it was. As for future expenditure, the Council agreed to set up a working party chaired by Bannier, one of whose main tasks was to propose suitable annual rates of growth for the CERN budget. If Britain had achieved a 'victory' in getting a ceiling policy accepted by the council in 1957, it had suffered a major 'defeat' in trying to exert a stricter form of external financial control over CERN in 1961.

SOME REASONS FOR THE BRITISH TREASURY'S BEHAVIOUR

In conclusion, I want to step back a little from these events and try to understand why it was that, at the end of 1961, the Treasury, aided and abetted by the Foreign Office, decided to try to take matters into its own hands regarding the financing of CERN. My explanations are at two levels. Firstly, I want to draw attention to a conflict *inside* the government itself. Secondly, I shall remark on the *external*, contextual situation of the day, contrasting certain aspects of the political and economic situations in Britain with those prevailing in France at the time.

If we begin at the micro-level, and ask why it was that the Foreign Office took the unprecedented step of proposing that CERN budget envelopes be agreed between governments, the answer is pretty obvious: the move was symptomatic of the Treasury's growing frustration at the British delegation's apparent inability to curb rising expenditure at CERN. That 'inability' was in fact indicative of an important interdepartmental disagreement within Her Majesty's Government. If the Treasury thought that the DSIR was weak and vacillating, some DSIR officials, in their turn, felt that the Treasury's attitude was not only mean but also diplomatically disastrous. It was 'obsessed by [Britain's] economic stagnation' and it 'resent[ed] the cost of [CERN's] success' wrote one DSIR official acrimoniously. And, as far as he was concerned, the results were entirely counterproductive. 'The usual British tactics over the past year or so', he wrote on another occasion, 'have been to declare far too conservative a policy to command support, to die in the last ditch in defence of our declared policy, and to see the moderates waver and, deprived of our leadership, go over to support the expansionists'.[11]

These attitudes were shared by the British CERN Council delegates, who were in the front line of battle as it were. Typically, Sir Harry Melville, who had this job at the time, actually welcomed the fact that matters were coming to a head at the end of 1961. Britain, he told the Council, had made a 'fatal' mistake when, in 1959, it had imposed an 'absolutely flat ceiling' on CERN's expenditure for three years – what was needed was an (upwardly) 'sloping ceiling'. That granted, as far as Sir Harry was concerned, the question now was: 'Are people going to support CERN as it ought to be supported or are they not?' British representatives could not come to Council meetings year after year with consistently lower figures than the others: this was 'not a satisfactory way of collaborating in international organizations' and 'the sooner we face up to this the better'.[12] In short, it was not 'Britain' that suffered a setback in the budget debate at CERN in 1961: it was the British Treasury and the policies that it held dear.

Why did the Treasury hold those policies? Three reasons are particularly pertinent. Firstly, there were the effects of the United Kingdom's 'special relationship' with the United States, here expressed as a doubt as to whether it was necessary to build up CERN to a level comparable to the best American facilities. The contrast with the French is striking here. The French, and the Gaullists in particular, generally disliked 'Anglo-Saxons', and were determined to be independent of American influence. Against this background it is not surprising to find a UK Treasury official wondering whether it was necessary for CERN to keep abreast of similar centres in America – or to see the French diplomat de Rose determined to build CERN into a laboratory which could 'keep up in the competition with big laboratories in the United States and the Soviet Union'.[13]

Economic considerations of two kinds also played a role. On the one hand, there was the state of the economy as a whole. As one commentator has put it, 'The postwar performance of Britain's economy has entered the annals of history as a tale of woe, of constant balance of payment crises [of which there was one in 1961], of overspending and living on credit.'[14] Related to this there was the level of investment in Research and Development, which advanced sluggishly in parallel: already higher in Britain than in any other European country (as a percentage of GNP) it only grew from 2.2 to about 2.4 per cent between 1962 and 1967.[15] The contrast with France is again illuminating. Her economy grew extremely rapidly after the general recession in Europe in 1957/8 – at between 4 to 7 per cent per annum between 1959 and 1963. At the same time, the Gaullist government also increased its R&D investment enormously – from about 1.5 per cent of GNP in 1962 to over 2.1 per cent in 1967, the steepest rise in this period of any country in the world.[16] Knowing this, we can understand why the Treasury baulked at the demands coming from a range of European joint enterprises, some of them entirely new (ESRO and ELDO), some of them well established, but exceeding all previous expectations (CERN), while de Rose had no hesitation in believing that all should be developed in parallel, and was totally unwilling to see CERN 'sacrificed in favour of new organizations', as the Foreign Office had implied in its *aide-mémoire.*

To conclude, one can say that Britain's policy at CERN was of a piece with its policy on European collaboration as a whole, subject to the same doubts and hesitations, shaped by the same overall financial and political considerations. From the start, much of the government and the élite of the British physics community were suspicious of the role being played by the French in pushing for CERN, feared that the laboratory's costs would spiral out of their control, were not sure whether the facility was really necessary scientifically, were not willing to sacrifice their domestic accelerator programme to it, and suspected that the same benefits could be derived more simply and cheaply by collaborating with the United States. These themes were to remain the *leitmotif* of Britain's policy on CERN for at least the next decade. Many of them continue to recur, even as the nation seems to be moving closer to believing that its future lies in Europe, with all the costs and benefits that that political choice entails.

NOTES AND REFERENCES

1. The primary source material on which this chapter is based is to be found in the CERN archives, Geneva, and in the archives of the Science and Engineering Research Council in Hayes, Middlesex.
   The chapter highlights Britain's dilemmas in the negotiations over the CERN budget which are described far more extensively in my Chapter 10 on finance policy in A. Hermann, J. Krige, U. Mersits, and D. Pestre, *History of CERN. Vol. 2. Building and Running the Laboratory* (Amsterdam: North Holland, 1990). This chapter contains a comprehensive list of references to the primary sources, so that I have decided to restrict the references given here primarily to important quotations. The interested reader is referred to the more extensive version for more detailed information.

2. We are fortunate in having a verbatim record of Verry's statement of the Finance Committee at its meeting on 1–2 May 1957 (attached to letter CERN/3748, 16.5.57, from Eliane Bertrand to the British delegate, and in file DG20804 in the CERN archives), in addition to the usual official minutes, document CERN/FC/198, dated 10.5.57. All of the quotations from Verry are to be found in these two documents.
3. More information on the French, Italian (and German) reasons for joining CERN, and a contrast with British arguments, can be found in Part IV of A. Hermann, J. Krige, U. Mersits and D. Pestre, *History of CERN. Vol. 1. Launching the European Organization for Nuclear Research* (Amsterdam: North Holland, 1987). See also J. Krige, 'Why did Britain Join CERN?', in D. Gooding, T. Pinch, and S. Schaffer (eds), *The Uses of Experiment* (Cambridge University Press, 1989) pp.385–406.
4. The contributions made by the SERC to CERN may be found in the Council's Annual Reports. The figure quoted here is that given for the International Contributions made by the Nuclear Physics Board in the year 1987/8.
5. For more information on the cohesion of the CERN Council see D. Pestre's Chapter 7 in *History of CERN. Vol 2,* cited in note 1. The quotations from Bannier are from the minutes of the FC meeting held on 16.10.57, document CERN/FC/242, 30.10.57, and those of the meeting held on 13.11.57, document CERN/FC/252, 22.11.57.
6. For the Foreign Office's *aide-mémoire*, see the circular from its Atomic Energy and Disarmament Department to the Chanceries in the CERN member states, dated 10.11.61, and in Box B125 in the SERC archives. For a Dutch reaction see the reply from the Ministry of Foreign Affairs in The Hague, dated 18.12.61, and in the same box.
7. Bannier made this remark at a meeting of the Committee of Council held on 18.12.61. The proceedings of the meeting from which the quotation is taken are document CERN/CC/435, dated 23.1.62; a tape recording is also available in the CERN archives.
8. The British contested de Rose's figures for the rate of growth. They argued that, since they were prepared to support a budget of SF75m for 1962, their SF240m envelope amounted to giving CERN SF80m for 1963 and SF85m for 1964 – an annual growth rate of over 6 per cent. De Rose disputed this. Starting on the assumption that the majority would vote the 1962 budget at SF78m, he said that the British envelope would mean that CERN got SF80m for 1963 and SF82m for 1964, so about a 2.5 per cent increase per annum.
9. The quotation from de Rose is from the minutes of the 20th session of the CERN Council, 19.12.61, at p. 37.
10. De Rose threatened to break off negotiations with Switzerland over the extension of the CERN site into France in an off-the-record remark to the Committee of Council at its meeting on 18.12.61. It can be heard on the tape recording of the meeting which is in the CERN archives, and to which we have very kindly been given access.
11. The criticisms of the Treasury were voiced by DSIR official G. Hubbard, the first in a memo to the DSIR's Secretary on 22.10.61, the

second in a memo to Elkington, 3.11.61. Both are in Box B125 in the SERC archives. Hubbard was an adviser to the official British delegation in the CERN Council.

12. Melville's remarks can be found on the tape recording of the Committee of Council meeting held on 12.10.61, which is in the CERN archives.
13. For Treasury doubts on the need to keep abreast of the Americans, see letter Adams to Griffiths, 19.1.62, Box B125 in the SERC archives. De Rose's remarks were made at the meeting of the Committee of Council on 18.12.61; see note 7.
14. The quotation is from W. Laqueur, *Europe Since Hitler* (Penguin, 1982)p.224.
15. Britain's figures for gross expenditure on R&D as a percentage of GNP in 1962 are taken from C. Freeman and A. Young, *The Research and Development Effort* (Paris: OECD, 1965) p.71. The figures for later years are taken from *OECD Science and Technology Indicators* (Paris:OECD, 1984) p.27.
16. The growth rates of France's economy are from Laqueur (see note 14) p.214. Figures for its R&D effort are from the sources quoted in the previous note.

7 European Countries in Science-Based Competition: The Case of Biotechnology

MARGARET SHARP

The purpose of this chapter is to consider how best Europe can maximise its advantages in biotechnology. It is wrong to talk about biotechnology as an industry. It is not an industry but a technology, or rather a set of technologies. It is above all concerned with new approaches to old problems and new ways of doing things; it is a process technology, not a product technology. (This helps to explain why it has led to so few new products on the market some 15 years after it first emerged.) To maximise its advantage in biotechnology, Europe's goal must be the rapid diffusion of the new process techniques. This chapter argues that the key to this diffusion process lies, not, as many would suggest, with the small firm sector, but with Europe's major pharmaceutical and chemical companies. It further argues that the best way to promote the take-up of biotechnological processes lies in strong support for the science base, the building of bridging mechanisms between industry and academia, and a tough but sympathetic regulatory environment.

WHAT IS BIOTECHNOLOGY?

There are many definitions of biotechnology. American definitions tend to stress genetic engineering, whereas in Europe emphasis has always been put upon the broader processing base from which it derives. Hence the OECD definition: 'the application of scientific and engineering principles to the processing of materials by biological agents' (Bull, Holt and Lilly, 1982). As such it encompasses the use and application of a series of skills ranging from molecular biology, biochemistry, genetics and microbiology to biochemical engineering and separations processing.

The case for emphasising that biotechnology emerged from earlier developments in genetics and fermentation technology is that it highlights the extent to which developments in genetic engineering in the early 1970s constituted a radical change of technological trends. Two seminal break-throughs – those leading to recombinant DNA (r DNA) and the cell fusion methodology of monoclonal antibodies – have opened up hitherto undreamed of developments in science, particularly in the fields of medicine and agriculture. Using the analogy of the Kuhnian scientific paradigm, Dosi (1982) developed the concept of the technological paradigm. He defines this as 'both a set of exemplars – basic artefacts which are to be improved (e.g. a car, a lathe, an integrated circuit) – and a set of heuristics' (Dosi, 1988, p.224). We may interpret his use of the word 'heuristics' to mean a basic set of ground rules – such as the principles of the internal combustion engine – which dictate the 'thought set' of those working in the field. It then becomes clear that genetic engineering provides precisely such a paradigm change in biotechnology. It genuinely has changed the whole set of basic ground rules and opened up new approaches to old problems, and new fields of development.

This is well illustrated in two fields where development has been most active – pharmaceuticals and agriculture. In pharmaceuticals, there are as yet few new products which derive from biotechnology and, as cynics are fond of telling us, many of the new drugs which will eventually emerge will still be based upon chemical synthesis. But their origin and the routes by which they will have been developed will owe little to the hit-and-miss chemistry of the 1950s and 1960s – the so-called golden age of pharmaceuticals – and a great deal of techniques such as protein engineering which have been developed with the new biotechnology. Indeed, by analogy with the successive generations of computers and chips, some observers have already identified three generations of 'the new biotechnology'. The first was associated with the cloning and expression of simple proteins such as insulin and the interferons; the second, and current, phase comprises attempts to use protein engineering to alter the characteristics and composition of such proteins; and the third, which is seen as emerging towards the end of the 1990s, will be the age of 'designer drugs' – slim-line proteins which can be designed, synthesised and used against specific ailments. In agriculture too, the genetic manipulation of plants has made possible the rapid development of new and selected plant species of a type, and on a time scale, which were impossible under existing methods of cross-fertilisation and selective breeding. Once again, biotechnology embodies a wholly different *approach* to plant breeding.

Three important characteristics of new biotechnology which are associated with this paradigm shift can be identified:

1. *Pervasive uncertainty.* The paradigm shift means that the techniques are, at least initially, untried and experimental, and that public reaction to the new, genetically-engineered products is unknown. Added to this, there has been great confusion over the patentability of both processes and products. Hence those moving in early to develop these techniques were faced by genuine uncertainty, with the risks not quantifiable, as to whether the new techniques would work, or, if they did work, whether the new products would be acceptable to the general public. By early 1982 it was clear that the first phase – the cloning and expression of simple proteins – had been mastered, and some of the patentability issues were resolved. But the fast movement of the scientific base, the problems of scaling up, and the remaining question of public acceptability left many issues still subject to considerable uncertainty.
2. *The importance of the academic science base.* More than in any other new technology (except perhaps nuclear power in the 1950s), the locus of knowledge in this developing field lay in the academic sector, and required mechanisms of technology transfer which both gave access to that knowledge base and helped to translate science into technology. In the process, the traditional boundaries between basic and applied science (and between science and technology) became blurred. What is more, the continued fast movement in the basic science base, for example developments in protein engineering, has meant continued dependence on academic research. In other new technologies, by contrast, relevant knowledge rapidly moved into the research laboratories of the company sector.

3. *The need to mix established disciplines.* Biotechnology cuts across the traditional skills of the pharmaceutical and chemical industries, just as it cuts across the traditional academic departments of most universities. In both academic and industrial settings it demands project teams which span traditional disciplines – protein chemists and biochemists, molecular geneticists and molecular modellers, microbiologists, protein crystallographers, fermentation scientists and separation process experts. For many firms it mean getting access to new skills and building up new cross-disciplinary, in-house capabilities. Putting teams together containing such diverse talents takes time and care. The short-term solution has been to buy-in the relevant skills on a temporary basis from the academic sector (where they have originated) or via the new small biotechnology firms.

THE NEW BIOTECHNOLOGY FIRM – AN AMERICAN PHENOMENON

The most publicised phenomenon in the development of biotechnology has been the emergence of the new specialist biotechnology firm, the NBF. This has above all been an American phenomenon and the story is well known. Starting with the launch of Genentech in 1976, the number of businesses grew rapidly from 10 in 1978 to 40 in 1980, to 85 by 1981, 150 by 1983 and to 250 by 1985 (OTA, 1984; *The Economist*, 1989). In spite of the fact that many firms fell by the wayside, both in the stock market crash of 1987 and in the normal rough and tumble of small-time corporate life, company start-up has continued. Today there are some 400 small biotechnology firms in the United States. As *The Economist* commented recently:

> Call it a miracle. Or maybe madness. Over the past decade hundreds of tiny biotechnology firms have sprung up and new ones are still being created almost every week ... Only one firm – Genentech – makes a sustained profit, and even that is disappointingly small. Yet investors continue to pour money into these companies on an heroic scale – $10 billion so far. How much longer can this industry defy gravity? (*The Economist*, 1989)

It is easy to see in retrospect why the NBF grew and flourished in the early years of biotechnology. Essentially it fulfilled two functions. It provided:

1. *A hedge against uncertainty*. The NBF can be seen as a very clever means by which larger firms were able to 'keep a window' on the technology while holding their own commitment (of money and people) to the minimum. Only after the uncertainties of the early years were eliminated did the big firms begin to commit themselves to substantial investments in building up their in-house capabilities.
2. *A bridge between academia and industry*. The prime function of the NBF was to provide a means of getting access to the scarce skills of the academic sector and puting them at the disposal of industry. In fact many NBFs were effectively private research laboratories for bright professors. They had the advantage of separating academic from commercial activities (which often suited both the university and the client); of enabling academics to realise the (financial) value of scarce skills; and of sharing these skills between clients. They also enabled project teams to be rapidly assembled across the various disciplines without upsetting established institutional frameworks, either within the academic sector or within industry.

Funding for the NBFs came partly from the large companies, some of which established equity partnerships with them; others took substantial equity stakes and many placed research contracts with them. But most funding came, as the quotation from *The Economist* implies, not from the large firms but from the venture capital market – from individuals and institutions who were prepared, in effect, to gamble on the development of particular NBFs.

Table 7.1 presents estimates derived from a United Kingdom government source of the number of biotechnology firms in the United Kingdom, France and West Germany compared to those in the United States. It accords with the earlier figure (derived from *The Economist*) of approximately 400 such firms in the United States, but other estimates give higher figures – approximately 1000 small firms in the United States (Sapienza, 1989)[1]), over 100 in the United Kingdom (*Financial Times*, 1988) and over 20 in France (Raugel, 1986). This variation underlines the difficulty of defining the boundaries of biotechnology. Many of the firms quoted, for example, in the

TABLE 7.1 *Biotechnology companies and sector of interest, 1986*

	US	*UK*	*France*	*FDR*
Agriculture	73	15	5	2
Chemical	37	4	1	4
Diagnostics	141	10	3	6
Food	18	12	2	1
Pharmaceuticals	65	9	2	4
Veterinary	54	6	3	0
Total	388	56	16	17

Source: Coleman (1987)

Directory of British Biotechnology (AABB, 1988) are specialist suppliers of instruments or equipment and have little or nothing to do with genetic engineering or biological processing. What is also clear is that, in all countries, including the United States, the number of small firms operating at the leading edge of technology – firms such as Genentech, Cetus, AmGen or Calgene in the United States; Celltech and British Biotechnology in the United Kingdom; Transgène in France – is small in relation to the total. Most firms are operating in niche markets as specialist suppliers of enzymes, assays or similar speciality goods, or as contract service firms to the wider biotechnology community. But, while Table 7.1 may underestimate numbers and/or give a false impression of the level of activity, the relativities remain correct. The small-firm sector has been, and remains, very active in the United States; is of growing importance in the United Kingdom; but has negligible presence in other European countries.

Why should this be? Why has the new biotechnology firm flourished in the United States but not in Europe? The answer lies partly in instutions, partly in culture, and partly in the quality of the academic base.

Institutions

One of the important institutional differences between the United States and Europe has been the absence in Europe of a developed venture capital market. The venture capital market in the United States effectively created the new biotechnology firm. For it was the venture capitalists who identified the key academic scientists, offered to set them up in what were to all intents and purposes private research laboratories, sold their services on contract to the large pharmaceutical and chemical firms, and then over-sold them to the general public via the flotation of public share offering. The process of 'hype' then became self-perpetuating, since it was necessary to keep shareholders happy by 'talking up' share prices. The ample availability of venture capital, and the copycat phenomenon in relation to the small electronics firm (with everyone by the late 1970s asking 'what's next?'), thus combined, as *The Economist* put it, to 'defy the laws of gravity'. They also led, however, to the premature commercialisation of biotechnology. Such were the uncertainties and the unpredictabilities of its early days that, in normal circumstances, many advances would have been regarded as experiments to be carried out in academic research laboratories, not yet mature enough to be transferred to the commercial sector. But the American venture capitalists were in too much of a hurry, the American public too gullible and American academics too ready to connive with the process, for the normal gestation in academic laboratories to be allowed to occur. Hence many of the delays and the disappointments, not only of the early years, but even today.

By contrast to the United States, the venture capital market in Europe is underdeveloped. The most active venture capital market is in the United Kingdom, where some half-dozen funds specialising in investment in biotechnology are active and an estimated total of over $1 billion has been invested since 1980 (*Financial Times*, 1988). The doyen of this market is the Rothschild Fund – Biotechnology Investments Ltd (BIL) – now capitalised at $200 million and the largest specialist fund in Europe. By contrast, the largest German venture capital fund, Techno Venture Management, established in 1984, had an initial capitalisation of $10 million (Yuan, 1987) and in 1989 was worth $50 million. The availability of venture capital is only one part of the equation, however. BIL, for example, whose investments span biotechnology and medical technology, have not found in Europe the quality of investment they are seeking. A total of 75 per cent of their investments are in the United States, only 25 per cent in Europe, and these concentrated almost entirely in the United Kingdom. This pattern of investment is mirrored by nearly all the investment funds in Europe, all of which direct a large proportion of their investments in biotechnology to the small-firm sector in the United States, and only a very small proportion to small firms in Europe.

Cultural and Institutional Factors

We need, therefore, to look beyond the availability of venture capital to find an explanation for the absence of a sizeable small-firm sector in Europe. An explanation frequently mentioned is the different academic ethos in Europe from that in the United States. In fact there are two different factors at work here. The first is cultural, the second institutional.

There remains in Europe, particularly in West Germany[2], an elitism which, consciously or unconsciously, puts curiosity-led academic science on a pinnacle and shuns commercial involvement or commercial interests. This tradition rejects out of hand any notion of academics playing the dual role seen in the new biotechnology firms in the United States. This is partly because of fears of conflicts of interest, but above all because such involvement is perceived to bias the development of academic science unduly towards commercial interest and away from advancing knowledge for its own sake. Often this elitism takes the form not of a positive shunning of commercial interests, but of a sublime indifference to and fundamental ignorance of the commercial world – not a good forcing ground for the new biotechnology firm. American science is not without elitism, but the dominant culture, encouraged by a university system in which many pay their way through college by taking on a variety of part-time jobs, is to cash in on opportunities which present themselves.

Allied to these cultural differences are the major institutional differences in the organisation of academic science in European countries. The United Kingdom comes closest to the US model, with medical research organised (via the Medical Research Council – MRC) in a mixture of institutes (e.g. LMB – Laboratory of Molecular Biology at Cambridge) and university-based programmes, but with the universities playing a substantial role in research both in the medical and the biological sciences. In West Germany, the universities continue to play the dual role of research and teaching base, but much of the highest-quality research is concentrated in the Max Planck Institutes, which are separate research institutes, often associated with a university or group of universities. Applied research is likewise concentrated in government-funded laboratories such as the GBF (Gesellschaft für Biotechnologie Forschung) at Braunschweig. In France, the process goes one step further. The universities rank lowest among higher education establishments and are mainly teaching institutions. The *grandes écoles* are the premier higher education establishments, but nevertheless do little research. Most research is undertaken in the publicly-funded research laboratories of the CNRS (Centre National de Recherche Scientifique), INSERM (Institut National de Santé et Recherche Médicale), and INRA (Institut National de Recherche Agronomique) or in the privately-financed institutes. Chief of these, for biotechnology, is the Institut Pasteur, which has provided the powerhouse for much of French work in the biosciences.

These differences in the organisation of scientific research may well be a major factor in explaining the small numbers of new biotechnology firms in Europe. There is first and foremost a straight conflict of interest. A researcher at a CNRS laboratory in France, or at Max Planck Institute laboratory in Germany, is the full-time employee of that institution. As such, his or her prime responsibility is to public, not private science. Moreover, as a full time employee, he or she will not find it easy to undertake the 'mix' of research frequently undertaken by an American professor, who combines an academic post with consultancy in the private sector. Indeed the tradition of funding US academic posts for only nine months of the year, expecting the academic who wishes to carry out research in the summer to raise research funds to meet the remaining three months of salary, explicitly encourages the entrepreneurial academic. In stark contrast, his or her German opposite number at a Max Planck Institute will find all research costs, including staff and equipment, met as part of institutional overheads. The opportunity cost of leaving such a research environment for the insecurity of the small firm is all the greater since, once off the academic ladder in West Germany, it is more difficult to climb back on again. The same goes for their opposite number in France, and with the additional disincentive that French researchers are civil servants, and dropping out of the system means *both* losing security of tenure or accumulated benefits *and* difficulty in re-entry, should the need arise. In the circumstances, it is not perhaps so surprising that few spin-offs from public-sector research arise; nor, for that matter, that in Europe most spin-offs are to be found in the United Kingdom, where the organisation of academic science most closely matches that of the United States. In the United Kingdom, it is notable that, with the exception of Celltech and the Agricultural Genetics Company (AGC),[3] most of the spin-offs from biotechnology have come from the universities.

The Science Base

Finally, we may look at the science base underlying biotechnology. Table 7.2, which is taken from work recently completed at SPRU by John Irvine, Ben Martin and Phoebe Isard, compares spending in 1987 on the biomedical sciences in the United States, France, Japan, the Netherlands, West Germany and the United Kingdom. The figures highlight the degree to which American support for the National Institutes of Health has fuelled the biotechnology revolution, with expenditure on the life sciences way above that of other

TABLE 7.2 *Breakdown of national expenditures on academic and related research by main field, 1987* [a]

	Expenditure (1987 $m)						
	UK	*FDR*	*France*	*Neths*	*US*	*Japan*	*Average*[b]
Engineering	436 15.6%	505 12.5%	359 11.2%	112 11.7%	1 966 13.2%	809 21.6%	14.3%
Physical sciences	565 20.2%	1 015 25.1%	955 29.7%	208 21.7%	2 325 15.6%	543 14.5%	21.2%
Environmental sciences	188 6.7%	183 4.5%	172 5.3%	27 2.8%	859 5.8%	136 3.7%	4.8%
Maths and computing	209 7.5%	156 3.9%	175 5.4%	34 3.5%	596 4.0%	88 2.3%	4.4%
Life sciences	864 30.9%	1 483 36.7%	1 116 34.7%	313 32.7%	7 285 48.9%	1 261 33.7%	36.3%
Social sciences (& psychology)	187 6.7%	210 5.2%	146 4.6%	99 10.4%	754 5.1%	145 3.9%	6.0%
Professional and vocational	161 5.7%	203 5.0%	67 2.1%	82 8.5%	490 3.3%	369 9.9%	5.8%
Arts and humanities	184 6.6%	251 6.2%	218 6.8%	83 8.6%	411 2.8%	358 9.6%	6.8%
Multidisciplinary	6 0.2%	32 0.8%	3 0.1%	1 0.1%	217 1.5%	28 0.8%	0.6%
TOTAL	2 798	4 037	3 212	958	14 904	3 736	

Note: [a]Expenditure data are based on OECD 'purchasing power parities' for 1987, calculated in early 1989.
[b]This represents an unweighted average for the six countries (i.e. national figures have not been weighted to take into account the differing sizes of countries).

Source: Irvine, J., Martin, B. and Isard, P., *Investing in the Future: An International Comparison of Government Support for Academic and Related Research* (Aldershot: Edward Elgar, 1990).

countries in both absolute and proportional terms. It also shows that the proportion of research monies spent on the biomedical sciences varies less across the three European countries, but the UK proportion is significantly lower than in West Germany or France.[4] In absolute terms, because its civilian research budget is so much higher than the United Kingdom, West German spending is almost twice as high. (The differences are less significant in manpower terms, because academic/research salaries are so much lower in the United Kingdom).

In spite of budgetary stringency, however, the United Kingdom continues to maintain its disproportionate contribution to this field of science. Table 7.3 identifies the contributions of five major countries to scientific literature in the field of biotechnology, using citation

TABLE 7.3 *Citations of scientific publications in fields relevant to biotechnology (country shares)*

		USA	*Japan*	*UK*	*FDR*	*France*
Biomedical	1973	55.3	3.0	11.2	4.8	4.0
	1984	55.1	6.6	9.4	5.7	4.2
Biology	1973	50.4	4.4	14.1	3.2	2.5
	1984	43.5	7.4	12.8	4.8	2.2
Genetics and heredity	1973	42.5	4.2	14.5	7.0	3.8
	1984	39.6	7.0	10.0	6.4	4.1
Biochemistry and molecular biology	1973	54.7	3.9	10.2	5.1	4.2
	1984	49.1	10.1	9.7	5.3	4.9
Biophysics	1973	50.8	0.2	8.6	6.3	0.9
	1984	72.1	1.8	7.4	4.1	0.2
Cell biology	1973	54.4	2.8	8.5	6.8	3.3
	1984	55.7	5.2	7.2	7.3	2.5
Microbiology	1973	52.7	4.1	15.1	4.0	3.1
	1984	39.8	4.7	17.4	8.2	4.7
Virology	1973	63.5	3.2	9.2	2.4	1.5
	1984	62.3	7.7	8.5	6.1	3.4

Source: CHI/NSF Data Base (held at SPRU, University of Sussex); adapted from Tanaka (1988).

data. The assumption here is that a much-cited paper will, in general, be a more influential one. It will be noted that, although the United Kingdom lost ground over the period 1973–84, its citations remained in most fields at least 50 per cent above the West German level, and were more than double those of the French. Again, the dominance of the United States is overwhelming. If, as was suggested, the small biotechnology firm is an important means of obtaining access to academic science, it is apparent why the US small biotechnology firm has so much more to offer than its European counterparts – and why the United Kingdom has more to offer than its French and German partners.

THE IMPORTANCE OF EUROPE'S LARGE CHEMICAL AND PHARMACEUTICAL COMPANIES

The conclusion to be drawn from the previous section must be that, although the small-firm sector can be expected to grow in Europe, it is not yet, and is not likely to be, an important agent for the diffusion of this new technology. The onus for diffusion is therefore upon the major chemical and pharmaceutical companies and on the technological transfer mechanisms linking these companies with the academic sector. Table 7.4 lists the ten largest pharmaceutical/chemical companies in Europe, together with their sales and R&D data, comparing them with similar companies in the US and Japan. These companies already have substantial involvement in biotechnology, although the depth of that involvement varies from company to company and, within companies, from sector to sector. As might be expected, however, all the companies have some involvement both in pharmaceuticals and in agriculture.

All these companies are also major international businesses, with extensive interests both in manufacturing and R&D in other continents. All have research laboratories in the United States; some have several research laboratories in that country. Many are establishing research facilities in Japan. As multinationals they have the ability to gain access to knowledge on a world-wide basis, and this is precisely what they have been doing. Ciba-Geigy, for example, in addition to its Swiss research facilities, which are extensive and include the Miescher Research Institute for the biological sciences in Basel, has research centres in North Carolina and in the United Kingdom. It also has two joint ventures in biotechnology, one with the NBF

TABLE 7.4 *Sales and R&D expenditure on the major chemical firms in Europe, USA and Japan*

	R&D expenditure 1987/8 ($m)	*Annual sales 1987/8 ($m)*	*% R&D/sales*
Bayer	1276	20635	6.20
BASF	896	22354	4
Hoechst	1231	20531	6
ICI	641	18242	4.10
Rhône Poulenc	585	9344	6.30
Ciba Geigy	1122	10580	10.60
Sandoz	540	6026	9
Akzo	451	8631	4.70
Montedison	291	10641	2.73
Enichem	158	—	3.60
Du Pont	1223	30468	4
Dow Chemical	670	13377	5
Union Carbide	159	6914	2.30
Monsanto	557	7639	7.30
Sumitomo Chem	88	5034	3.19
Ashai Chem	234	6621	3.50
Takeda Chem	216	4378	4.90
Mitsubishi Ch	221	7586	5.10

Source: Annual reports, *European Chemical News.*

Chiron and one with Corning Glass; and it has research contracts/linkages with six other new biotehcnology firms in the United States and three in Europe. Bayer, which has developed close links with the Max Planck Institute for Plant Breeding in Cologne, established its own (internal) Institute for Biotechnology at Mannheim in 1983. It has three separate research centres in the United States – the Miles Cutter Laboratories at Elkhart, the Mobay (agricultural/veterinary) Research Center in Kansas, and the West Haven Research Center, established in 1985, which concentrates on molecular biology and immunology. Bayer also has close links with the Universities of Rochester and Yale, and has had a research link-up with Genentech to develop Factor VIII, a blood product used in the teatment of haemophiliacs.[5]

Although Bayer and Ciba-Geigy are amongst the more internationally-active of these major companies, their linkages illustrate how easily companies of this size, operating as they do as global players in the global market-place, can by-pass technological deficiences in one market and 'plug in' to another. This was most dramatically illustrated in the Hoechst deal with the Massachesetts General Hospital (MGH) in 1979/80. At that time the skills that Hoechst needed in genetic engineering were not available in West Germany: the linkage with MGH enabled them to train staff and learn techniques. (The recent renegotiation of the contract dealt with different issues, since the skills Hoechst needed in 1979 are now readily available in the firm, and more generally in West Germany.)[6] In this way, international companies are themselves acting as a technology transfer mechanism, transferring skills from one country to another, which raises interesting questions about 'laws' of comparative advantage. Clearly, the advantages of academic supremacy can be easily 'captured' by multinationals and Europe's competitiveness in biotechnology has benefited greatly from this exchange.

Interestingly, the companies with fewest links with the US NBFs – or indeed with the US academic sector in general – are UK companies, although ICI, Glaxo and Wellcome all have major research establishments in the United States. They also have close links with UK academic research and participate actively in a number of the collaborative programmes in biotechnology organised by the government. This may indicate a more cautious approach to biotechnology than that of many of their counterparts in other countries, or greater competence of the UK academic sector in these areas (and the satisfactory nature of the linkage between the academic sector and industry); it seems probable that both factors play a part.

CONCLUSIONS AND POLICY IMPLICATIONS

The policy issue posed at the beginning of this chapter was not how best Europe might compete in biotechnology but how it might maximise its advantages. This is important because the previous section suggests that Europe's ability to compete – in markets where biotechnology is likely to have a significant impact – depends primarily upon the diffusion of this new technology among its large chemical and pharmaceutical businesses. As international firms, with a presence in most major markets, they respond primarily to their perceptions of the strategy adopted by competitors. The diffusion of biotechnology depends on the pace at which these firms, and their counterparts, judge it appropriate to shift to the use of biotechnological processes. To date this has happened relatively slowly, in spite of all the initial hopes, with most of the pack keeping fairly close to one another. European-based multinationals are very much 'up with the pack' and, as we have seen, will use their international linkages to ensure that this remains the case.

Maximising Europe's advantage from biotechnology implies a somewhat different objective, that of maximising the value added from developing and using a new technology. By this criterion, it is clearly to Europe's advantage that the multinationals should choose to locate research and development deriving from the new technology in Europe rather than abroad, so that Europe may gain the maximum value added from developments.

The analysis of this chapter suggests a number of policy initiatives which would help Europe to maximise its value added from biotechnology.

1. *More support for basic science.* The magnet attracting international firms to investment in US research in biotechnology is the sheer quantity and quality of the US research base. European countries invest a smaller proportion of research funding in the life sciences than does the United States, and there seems to be room for expansion. On quality, there may be something to be learned from the United Kingdom, whose share of citations in this area is disproportionately high. But the basic lesson is that, if Europe wants to retain or attract a bigger proportion of international R&D in biotechnology, it needs to invest more in the bioscience infrastructure.
2. *Institutional arrangements matter.* The second lesson is that institutional arrangements for science matter. Research institutes may relieve the academic of the necessity to generate his own research funds and, in principle, make it easier to put together cross-disciplinary teams in areas such as biotechnology. But they clearly pose problems of technology transfer which have yet to be solved. There is some suggestion from the figures quoted that the US/UK pattern of academic organisation, with its emphasis on a mixed teaching/research environment and project/programme grants rather than the the establishment of tenured posts, is more conducive to academic creativity and to 'accessibility' by others, be it the broader academic community or the industrial community.
3. *Bridging mechanisms also count.* If Europe lacks the small-firm intermediary, then formal and informal bridging mechnisms between academic science and industry assume greater impor-

tance. There is now a wealth of experience of such mechanisms and a diversity of them across Europe. Clearly the institutional framework and the cultural background influence appropriate mechanisms, but there may be lessons to be learned from comparative experience. The United Kingdom has seen a particularly wide range of schemes promoted by the Research Councils and the Department of Trade and Industry, some of which have been more successful than others. In particular, collaborative clubs in biotechnology have attracted attention, and these could perhaps prove adaptable to the environments of other countries.

4. *The regulatory environment.* The regulatory environment assumes far greater importance for biotechnology than for other new technologies. Fears about the safety, controllability and the long-run impact of genetic engineering are understandably strong, so that the regulatory authorities need to establish codes of conduct which are adhered to and can be enforced. That said, there is now increasing experience and experimentation on which to base risk assessments and many of the early fears about the release of uncontrollable mutations have been quashed. At present, there is a diversity of codes and regulations from country to country, but these are gradually being 'harmonised' within a single regulatory framework for the European Community (EC). The stringent regulatory framework imposed in West Germany and Denmark is currently having a substantial adverse effect on the development of biotechnology in those two countries. BASF have already moved their biotechnology research from West Germany to the United States, and other companies, among them Hoechst and Bayer, are contemplating doing likewise. If the EC regulations are as tough (restrictive ?) as those currently in existence in West Germany, the European research base in biotechnology will substantially disintegrate. Both research and product manufacture will be shifted to countries with less stringent controls. This would have an immediate impact on the small and medium-sized firms making use of biotechnology, since much of the infrastructure on which they depend would collapse. In the longer run, it seems impossible that it would have no impact upon the competitiveness of the major European-based multinationals.
5. *Competition.* Many of the main European-based chemical and pharmaceutical companies have been relatively slow, as com-

pared to their US counterparts, to develop capabilities in biotechnology. The reason for this may lie in part in the virtual absence of a small-firm sector in Europe. Europe may have been less affected by the 'hype', but equally lacked the stimulus to innovation that an active small-firm sector provides. The fragmented regulatory framework in Europe also helps these major firms to insulate themselves from the full rigour of competitive markets. The development – after 1992 – of a single regulatory framework for pharmaceuticals and agricultural chemicals through the EC will bring new competitive pressures on many of these firms. This should be reinforced by a tough anti-trust policy (to ensure that the pressures are maintained) including measures to support the growth of the small-firm sector.

6. *Measures to promote the venture capital market.* Although it has been argued that promoting the small-firm sector will not, *per se*, help Europe compete in biotechnology, the presence of the small-firm sector is important both as a vehicle for new ideas and, from time to time, as a way of pricking the complacency of the large companies. For these companies to become, as they are in the United States, major intermediaries between academic science and industry requires institutional and cultural changes which cannot occur overnight. In the longer run, if the sorts of measure proposed in (2) above are implemented, the role of the small biotechnology firm in Europe may well become closer to that of these firms in the United States. In the meantime, a maturing venture capital market can only be an advantage.
7. *The long-term cost of capital.* This is an issue applicable to the promotion of all new technologies, not just biotechnology. The real long-term cost of capital is today at an historic high. Innovation and the diffusion of new technologies require the acceptance of long-term risks. In the past, the tradition in some European countries has been for the banking sector to encourage long-term thinking, rather than the short-term views that seem to dominate the stock markets of New York and London. The ability to take a long-term view – that is to act independently of others – derives partly from the fragmentation of financial markets. It is vital, with the deregulation of capital markets within the EC, that the long-term approach is not driven out by the quick profits engendered by short-termism.

NOTES

1. Sapienza (1989) quotes figures derived from a presentation given at Biotech 88 in California in the autumn of 1987 by the Arthur Young High Technology Group. These figures, like those of the *Directory of British Biotechnology* (AABB, 1989) include many firms which are on the periphery of biotechnology.
2. Interestingly, in spite of their tradition of close links with industry, German universities remain the most elitist in Europe, the main links in the past having been forged via the technological universities. This rigidity is now breaking down, but the process is hindered by a highly inflexible career system which limits the number of career posts available and makes it difficult to bring in new, young researchers who have experience outside the university system (see Sharp, 1985; Yuan, 1987).
3. Celltech was established in 1980 as a government-initiated small biotechnology company, expressly set up to help to promote the transfer of technology from the Medical Research Council Institutes. The AGC was established in 1984 as the 'country cousin' of Celltech, to help to commercialise the results of research by the Agriculture and Food Research Council in the United Kingdom (see Sharp, 1985).
4. In fact, when the first set of these figures was put together (ABRC/SPRU, 1986) the proportion going to the life sciences in the United Kingdom was 34.4 per cent – significantly higher.
5. I am indebted to Ilaria Galimberti, a doctoral student at SPRU, for the information contained in this paragraph.
6. This became apparent from recent discussions between the author and senior research personnel from Hoechst.

REFERENCES

AABB (1988) *Directory of British Biotechnology.*

ABRC/SPRU (1986) *International Comparisons of Government Funding of Academic and Academically Related Research* (obtainable in mimeo from SPRU, University of Sussex, Brighton, BN1 9RF, UK).

Bull, A. T., G. Holt and M. Lilly (1982) *Biotechnology: International Trends and Perspectives* Paris: OECD.

Coleman, R.F. (1987) 'National Policies and Programmes in Biotechnology'. paper presented at Canada–OECD Joint workshop on National Policies and Priorities in Biotechnology, Toronto, Canada, 7–10 April. Reproduced in OECD (1988) *Biotechnology: The Changing Role of Governemt* Paris: OECD (1988).

Dosi, G. (1982) 'Technological Paradigms and Technological Trajectories: A Suggested Interpretation of the Determinants and Directions of Technical Change', *Research Policy*, vol. 2, no. 3, pp 147–62.

Dosi, G. (1988) 'The Nature of the Innovative Process', Chapter 10 in G. Dosi, C. Freeman, R. Nelson, G. Silverberg, and L. Soete, (eds), *Technical Change and Economic Theory* (London: Frances Pinter).

The Economist (1989) *The Money Guzzling Genius of Biotechnology*, 13 May, pp.91–2.

Financial Times (1988) *Supplement on Biotechnology*, p.II 'Ventures that may appeal to the heart of the City', 28 May.

Irvine, J., Martin, B. and Isard, P., (1990) *Investing in the Future: An International Comparison of Government Support for Academic and Related Research* (Aldershot: Edward Elgar).

OTA (1984) *Commercial Biotechnology: An International Analysis* (US Congress Office of Technology Assessment, January).

Raugel, P.J. (1986) 'Specialised Biotechnology Firms Financed by Venture Capital', *Biofutur*, special issue on Biotechnology in France, March.

Sapienza, A. (1989). 'Technology Transfer: An Assessment of the Major Institutional Vehicles for Diffusion', *Technovation*, 9, Summer.

Sharp, M. (1985) *The New Biotechnology: European Governments in Search of a Strategy* (Sussex European paper No 15: Science Policy Research Unit). (University of Sussex:)

Tanaka, M. (1988) *Industry-University Relations in the case of the new Biotechnology in Japan,* mimeo, Saitama University. Tokyo, Japan

Yuan, R. (1987) *Biotechnology in Western Europe* (US Department of Commerce: International Trade Administration, April).

8 Overseas Funding for Industrial R&D in the United Kingdom

PAUL STONEMAN

INTRODUCTION

Discussions of R&D in the United Kingdom usually centre upon the inadequacy of total expenditure relative to that in other competing nations. A phenomenon that appears to have merited much less discussion is that highlighted in Table 8.1. Over the period since 1976, the data on the sources of funding for R&D in the UK industrial sector have as their most noticeable feature a dramatic increase in the percentage of this total funding that comes from overseas. In this chapter I explore this phenomenon in more detail and attempt to address two basic questions: why is the percentage of UK industrial R&D financed from overseas relatively large and what are the costs and benefits of this?

It may at first sight appear that these questions are not immediately relevant to the management of science. However this is not actually the case. I would argue this on two grounds. First, the increase we observe has been attributed to 'R&D performed in the UK by overseas-controlled enterprises' (Cabinet Office, 1988, p. 34), and thus the issue to be discussed relates to the important management question of where R&D should be undertaken. Secondly, the costs and benefits to the United Kingdom of overseas funding for industrial R&D are an important input into technology policy discussions of

TABLE 8.1 *Sources of funds for UK industrial R&D, 1967–87*

	Total industrial R&D (£m)	*Overseas funds (£m)*	*Overseas funds as % of total*
1967	611.5	23.3	4
1968	647.6	29.1	4
1969	693.9	32.6	5
1972	838.5	54.0	6
1975	1340.1	84.6	6
1978	2324.3	185.2	8
1981	3792.5	331.3	9
1983	4163.3	283.2	7
1985	5121.6	569.0	11
1986	5950.7	727.0	12
1987	6337.0	784.2	12

Source: 'Industrial R&D expenditure 1987', *British Business,* 3 Feb. 1989. p. 24.

government on (i) whether such funding should be encouraged and (ii) whether a policy initiative is necessary to attain for the United Kingdom greater benefits from such overseas-funded R&D.

Prior to moving on to some detailed discussion it is necessary first to clarify what the overseas funding comprises. Unfortunately this is not exactly clear. It will, of course, cover R&D performed in the United Kingdom by overseas-controlled enterprises, but only that part funded from overseas, not the part funded from UK subsidiaries of such enterprises. It would also cover contributions from overseas to joint venture R&D undertaken in the United Kingdom. In addition it should include contract R&D undertaken by a UK company for a foreign customer. Finally it should include R&D funded under, for example, EC technology incentive schemes, such as ESPRIT, where part (or all) of the cost of the R&D is paid by Brussels. It is therefore quite a mixture of financing. We should hold in mind, however, the Cabinet Office statement that most of the increase in such funding is R&D performed in the United Kingdom by overseas controlled enterprises, that is, transnational corporations. Data from *Business Monitor* MO14 illustrate that approximately 50 per cent of all overseas funding for R&D in the United Kingdom goes to overseas-controlled enterprises, and overseas-controlled enterprises undertake 18 per cent of all industrial R&D in the United Kingdom. We should also note that UK companies will undertake R&D overseas funded from the United Kingdom. We do not have any data on this, however.

In the following sections, I explore more fully the financing of R&D from overseas; the causes of the phenomenon are explained; the costs and benefits are discussed; and, finally, some conclusions are drawn.

THE PATTERN OF OVERSEAS FINANCE FOR INDUSTRIAL R&D

In Table 8.1 above the primary data that illustrate the phenomenon are presented. In Tables 8.2 and 8.3 some international comparisons of the phenomenon are presented. These data cover six OECD countries including the United Kingdom. Unfortunately data were not available for the United States. From Table 8.3 it is clear that the proportion of Industrial R&D funded from overseas is much higher in the United Kingdom than in the other five competing nations. In West Germany, Sweden and Japan, there are only minimal proportions of such funding, and, even compared to France, the UK proportion is almost twice as high. The data in Table 8.2 illustrate that the high proportion in the United Kingdom is not just the result of low total industrial R&D. The total funds coming from overseas to the United Kingdom in 1985 were 73 per cent higher than in the next-highest country, France, and between 1981 and 1985 grew twice as fast. In fact, in 1983, the United Kingdom received more funds for R&D from overseas than the other five countries together. The data in Table 8.2 are in constant 1980 prices and we may thus see that, between 1967 and 1985, funds from overseas for industrial R&D in the United Kingdom increased by a factor of four in real terms.

In Table 8.4 data on the sectoral breakdown of industrial R&D for 1967 and 1985 are presented. Looking first at the 1985 data, one can see that the overseas funds are concentrated in four sectors, Aerospace, Chemicals, Electronic Equipment and Components, and Office Machinery and Computers. These four sectors also show the highest ratios of industrial R&D being funded from abroad, although by this test Aerospace and Electronic Equipment are not as heavily dependent as the other two sectors. Comparing 1985 to 1967, we see that funds from abroad have increased for all sectors (except Contruction, where the figure is small anyway), but especially in Chemicals, Electronic Equipment and Components, and Office Machinery and Computers. Moreover in all industrial sectors (except Construction) the proportion of Industrial R&D funded from overseas is higher in 1985 than in 1967, and in some cases significantly so.

TABLE 8.2 *Business enterprise R&D (BERD) (natural sciences and engineering), in 1980 dollars (millions)*

	UK		*France*		*FRG*		*Italy*		*Japan*		*Sweden*	
	Total	*Funds from abroad*	*Total*	*Funds from abroad*	*Total*	*Funds from abroad*	*Total*	*Funds from abroad*	*Total*	*Funds from abroad*	*Total*	*Funds from abroad*
1967	5 102	197	3 491		4 607		1 032		3 913		663	2
1975	5 176	327	4 310		6 428		1 701		7 979		809	17
1978	5 975	477	4 738				1 671		9 129			
1979			5 008	343	8 807	196	1 883	40	10 306	9	943	12
1981	6 582	575	5 594	394	9 155	109	2 258	96	13 108	17	1 090	19
1983	6 186	687	5 926	271	9 747	137	2 511	107	16 042	23	1 264	22
1985	7 150	799	6 685	461			3 105	189	20 290	21	1 646	23
1986							3 586	214				

Source: OECD STIU data bank; *British Business*, 3 Feb. 1989.

TABLE 8.3 *Percentage of BERD (natural sciences and engineering) funded from abroad*

Year	*UK*	*France*	*FDR*	*Italy*	*Japan*	*Sweden*
1967	3.8					0.4
1975	6.3					2.1
1978	8.0					
1979		6.8	2.2	2.1	0.1	1.3
1981	9.0	7.0	1.2	4.3	0.1	1.8
1983	7.0	4.6	1.4	4.3	0.1	1.8
1985	11.0	6.9		6.1	0.1	1.4
1986	12.0			6.0		
1987	12.0					

Source: OECD STIU data bank; *British Business,* 3 Feb. 1989

We may thus summarise the picture as one where increased overseas finance is a general phenomenon, but it is most apparent in Chemicals, Electronic Equipment and Components, and Office Machinery and Computers, with a significant presence in Aerospace.

Given that UK industrial R&D is largely concentrated in these four sectors, is it the case that the proportion of industrial R&D in the United Kingdom funded from overseas is high because such funding is a characteristic of these four sectors? Consider the data in Table 8.5, which compares the United Kingdom to France and Italy (the most interesting comparison). As can be seen, the funds from abroad are greater in the United Kingdom than in France or Italy in all industrial sectors (except Aerospace and Food Drink & Tobacco compared to France), not just in the four major funded sectors. Moreover, although in Italy and France overseas funding is concentrated in the four sectors of Aerospace, Chemicals, Electronic Equipment and Components, and Office Machinery and Computers, as in the United Kingdom, the proportions of overseas funding in Italy and France (except for Aerospace) are lower than in the United Kingdom. We may thus argue that, although these four sectors are the main sectors in which funds from overseas finance domestic R&D, the high ratio of such funding to total industrial R&D in the United Kingdom is not just the result of the concentration of UK industrial R&D in these four sectors.

The overall picture is thus one where funding from overseas for industrial R&D in the United Kingdom is higher than in comparable economies and has been increasing faster than in other economies,

TABLE 8.4 *UK BERD (natural sciences and engineering) by industrial sector, in 1980 dollars (millions)*

	BERD total		*Funds from abroad*		*% of total funds from abroad*		*Funds from abroad as % of total*	
	1967	*1985*	*1967*	*1985*	*1967*	*1985*	*1967*	*1985*
ALL SECTORS	5 102	7 150	197	799	100	100	3.8	11.2
TOTAL MANUFACTURING	4 776	6 603	190	719	96.8	90	4.0	10.9
Aerospace	1 209	1 175	76	98	38.6	12.3	6.3	8.4
Chemicals	702	1 385	38	201	19.2	25.2	5.4	14.5
Construction	37	30	2	1	0.8	0.1	4.0	3.3
Electronic eqpt & components	787	1 927	18	151	9.3	18.9	2.3	7.8
Electrical machinery	258	190	3	8	1.5	1.0	1.1	4.3
Food, drink & tobacco	159	171	3	13	1.4	1.6	1.8	7.4
Ferrous metals	123	50	1	4	2.3	0.5	0.4	7.2
Machinery n.e.c.	450	365	5	25	2.7	3.2	1.2	7.0
Motor vehicles	376	516	1	26	0.4	3.2	0.2	5.0
Office machinery & computers	110	497	25	178	12.9	22.3	23.1	35.9
TOTAL SERVICES	286	401	6	9	3.0	1.1	2.1	2.2

nec– not elsewhere classified
Source: OECD STIU data bank.

TABLE 8.5 *BERD (natural sciences and engineering) by industrial sector, 1985, in 1980 dollars (millions) UK, France, Italy*

	BERD total			*Funds from abroad*			*Funds from abroad as % of total*		
	UK	*France*	*Italy*	*UK*	*France*	*Italy*	*UK*	*France*	*Italy*
ALL SECTORS	7 150	6 685	3 105	799	461	189	11.2	6.9	6.1
TOTAL MANUFACTURING	6 603	6 195	2 810	719	457	182	10.9	7.4	6.5
Aerospace	1 175	1 256	353	98	187	84	8.4	14.9	23.8
Chemicals	1 385	1 302	646	201	44	34	14.5	3.4	5.2
Construction	30	44	13	1	0	0	3.3	0.5	0.6
Electronic eqpt & components	1 927	1 417	433	151	54	29	7.8	3.8	6.7
Electrical machinery	190	230	228	8	0	1	4.3	0.1	0.4
Food, drink & tobacco	171	86	23	13	10	0	7.4	11.3	0.4
Ferrous metals	50	75	13	4	0	0	7.2	0.3	0
Machinery n.e.c.	365	247	162	25	11	0	7.0	4.6	0.1
Motor vehicles	516	684	415	26	2	0	5.0	0.3	0
Office machinery & computers	497	334	226	178	110	14	35.9	32.8	6.2
TOTAL SERVICES	401	375	285	9	1	7	2.2	0.4	2.3

Source: OECD STIU data bank.

but is concentrated in four main industrial sectors: Aerospace, Chemicals, Electronic Equipment and Components, and Office Machinery and Computers. These observations make analysis of this phenomenon particularly appropriate for the United Kingdom.

THE FACTORS BEHIND OVERSEAS FINANCE FOR R&D

Having illustrated the extent of overseas finance for UK industrial R&D, the next task is to ask why R&D should be undertaken in one country but funded from another. As has been stated, overseas funding covers a number of different types of finance, such as funding from international agencies and funding by overseas-controlled enterprises, that is, transnational corporations. As the latter are considered to be major providers of such finance, the issue we will concentrate upon in this section concerns why a foreign-owned multinational should finance R&D in the United Kingdom with overseas funds.

Although TNCs concentrate most of their R&D at home, there has been a growing tendency for research facilities to be located overseas. Hood and Young (1982) classify three types of overseas R&D activities by TNCs:

1. Support laboratories, which act basically as technical service centres.
2. Locally integrated R&D laboratories, concerned with local product innovation and development and the transfer of technology.
3. International interdependent R&D laboratories, acting as basic research centres which may or may not interact with the firms' local manufacturing affiliates, and producing innovations as part of the co-ordinated world R&D programme of the parent TNC.

It would seem apparent that types 1 and 2 would be closely related to the local production activities of a TNC, and one might thus expect the majority of the funding for such R&D (perhaps with the exception of some start-up finance) to come from the local subsidiary. Overseas finance for R&D is much more likely to be related to the third type of activity, which can be completely independent of any local production facility.

This third type of R&D activity is largely the province of what Behrman and Fischer (1980) call 'world market' firms, global corporations whose orientation is to world rather than national markets, and whose globally integrated production strategy is leading to the establishment of specially designed international interdependent research laboratories. Overseas finance for R&D in the United Kingdom is concentrated, as shown above, in four main sectors, Chemicals, Electronics, Computers and Aerospace. These sectors do in fact serve world markets, and thus the observed pattern of finance would be consistent with a large part of the overseas funding for R&D being of this third type.

If one accepts that this third type of R&D activity is a major use of overseas funds for R&D (although unfortunately we have no data to show it is so) what factors are most important in determining its location? Behrman and Fischer (1980) and Dicken (1986) both argue that the major locational criteria for such R&D activities would seem to be the availability of qualified scientists and engineers (QSEs) and access to sources of basic scientific and technical development. In addition, world-market firms consider the ability to gain access and contacts with foreign scientific and technical communities as important, and a main means of achieving this is through links with universities. Behrman and Fischer (1980) state: 'Every one of the world-market firms stressed the need for a strong local university system as a prerequisite for choosing an overseas location for R&D.' Taggart (1989) also considers that the low relative costs of R&D professionals overseas may be relevant.

In terms, then, of why the United Kingdom should be especially attractive for such R&D facilities, we may argue that three main factors will be important: (i) the supply of highly skilled manpower; (ii) a strong university system; and (iii) the lower relative costs of UK QSEs. There may also be a further relevant point. As Behrman and Fischer (1980) point out, most of the world market firms are US transnationals (the Japanese do not tend to locate research facilities overseas). This gives the United Kingdom a language-base advantage relative to other countries, and there may also be advantages arising from the strong links and similarities between US and UK universities.

Finally, before moving on to consider the costs and benefits of such activities, I would like to point out the similarity between the phenomenon we are discussing and the 'brain drain'. Basically, a brain drain occurs when qualified scientists and engineers move overseas to work for an overseas corporation. However, if the overseas corporation locates in the United Kingdom and employs

QSEs there but takes the results of their research overseas, the outcome is very much like a brain drain. the advantage to the overseas corporation, however, is that it can employ staff at the lower local pay rates and will not have to overcome problems of culture shock, or the family problems that might come from transferring labour overseas.

THE COSTS AND BENEFITS OF OVERSEAS-FINANCED RESEARCH IN THE UNITED KINGDOM

For the purposes of this section, we characterise typical overseas-funded research as being research of 'world-market' TNCs attracted by skilled labour, a strong research infrastructure and relatively low wage rates, and unrelated to production in the United Kingdom. Of course, as has been argued, not all overseas finance for R&D will be such, but it is a useful characterisation.

The potential benefits of such research to the United Kingdom will include:

1. the accumulation of skills and knowledge in the employed labour force that might leak out on re-employment to other sectors in the economy;
2. multiplier effects of various kinds in related industrial sectors that will provide inputs to the research facility;
3. a possible attractant for future manufacuring facilities of the parent TNC

Behrman and Fischer (1980) summarise such effects thus: 'All in all, the benefits [to a host country] resulting from the local performance of R&D by transnational corporations are rather non-specific in nature.'

What, then, are the costs of such activity? Given that the major resource being used is the skill base plus the research infrastructure of the host economy, if the price paid for such resources by the TNC does not reflect their opportunity costs or their replacement cost (which in an ideal world would be equal), then, in the absence of significant positive externalitites (which our list of benefits suggests do not exist), the TNC's activities would be detrimental to the host economy.

Economic theory suggests that, in a perfect market, resources would be valued at their opportunity and replacement cost. However the market (that is the supply and demand) of technological resources is generally considered most imperfect. The training and education of skilled manpower in the United Kingdom is heavily subsidised by the state. Also skilled engineers are often considered to be underpaid in the United Kingdom. Both observations suggest that the price to be paid for technological resources in the United Kingdom is below opportunity costs or replacement cost. The use of these resources by a TNC would thus represent a subsidy to the research of the TNC by the UK taxpayer. In the absence of significant returns on this, this would not appear to be in the interest of the United Kingdom.

An alternative view might be to consider that the employment of UK scientists and engineers by a TNC will increase the demand for such QSEs and drive up their price (if the market works) or reduce their availability (if non-price rationing prevails), thus increasing the cost or reducing the resources available for research by domestic firms. Assuming that the domestic firm's R&D will yield greater benefits to the United Kingdom than the research of a world-market TNC, this effect will not be a desirable one. If UK research generates production in, profit remission to, or licence fee remission to the United Kingdom, which is more likely if the research is not being undertaken by a world market TNC, then the United Kingdom will benefit more.

The prime question therefore concerns whether overseas-financed research is really just research exploiting underpriced technological resources in the United Kingdom with little spin-off to generate compensating benefits, or whether it will underlay and provide much greater technological activity, resources and production in the United Kingdom. We cannot give a definitive answer as to how the balance works out in practice, but one can at least draw attention to the potential dangers.

Finally, almost as an aside, we can point out how TNC funding from overseas of research in the United Kingdom can distort the picture of relative international technological performance. Often a country's technological performance is measured by its relative R&D performance, which is seen as the basis of its technological advantage in production relative to other countries. If however, that R&D, through a world-market TNC, supports production in another economy, comparative R&D statistics become very poor indicators of technological potential.

CONCLUSIONS

In this chapter the extent of overseas financing for industrial R&D in the United Kingdom has been shown to be significant, growing and greater than in comparable economies. This funding is concentrated in four main sectors, Chemicals, Electronics, Aerospace and Computers. It was argued that this funding could in some part be attributed to the establishment of research facilities in the United Kingdom by world-market-orientated transnational corporations, and may not be related to production there. The costs and benefits to the United Kingdom of such activity were discussed and their potential harmful effects were highlighted.

The remaining questions relate to policy. Should the UK government encourage or discourage such investment by TNCs? In the absence of definitive quantification of costs and benefits, this question cannot really be answered authoritatively. What one can say, however, is that in view of the potential costs to the United Kingdom of such activities, policies aimed at increasing domestic benefit from such research should at least be considered.

REFERENCES

Behrman, J. and W. Fischer (1980) *Overseas R&D Activities of Transnational Companies* (Cambridge, Mass: Oelgeschlager, Gunn & Hain).

Cabinet Office (1988) *Annual Review of Government Funded R&D, 1988* (London: HMSO).

Dicken, P. (1986) *Global Shift* (London: Harper & Row).

Hood, N. and S. Young (1982) 'US multinational R&D: corporate strategies and policy implications for the UK', *Multinational Business*, 2, pp.10–23.

Taggard, J.H. (1989) 'The Pharmaceutical Industry: Sending R&D Abroad'. *Multinational Business*, 1, pp10–15.

9 Public Understanding and the Management of Science

BRIAN WYNNE

Conventional analysis of science policy or the management of science pays little or no attention to the public dimension, in terms of public reactions to science and technology. There is a striking polarisation between the 'policy for science' interest in organising and directing science for national or industrial economic advantage, and 'science for policy', where concerns about persuading the public to accept 'expert' perspectives on environmental and risk issues, for example, have become paramount.[1] So far there has been no interest in asking whether public responses to 'science', for example in the form of expert pronouncements and prescriptions about risks, have anything to do with the ways that science is managed.

Polarisation also dominates debate about public participation in the management and control of scientific research, as opposed to participation in decisions about its technological and social effects. The language of exclusion of the public has shifted from the euphoric 'endless frontier' visions of self-managed scientific expansion in the postwar years to anxious modern appeals in certain fields to avoid 'mob rule' and the curtailment of scientific inquiry *per se*.[2]

In the wider sphere, away from particularly controversial areas, the scientific community is now concerned about lack of political support for independent research (that is, research funded from the public purse), not because of *antipathy* so much as *apathy* on the part of the general public, which appears not to have even noticed the crisis signalled by eminent bodies such as 'Save British Science'. This lack of expressed public interest in supporting science may be described as a *legitimation vacuum*, in the sense that the public seems unable to appreciate the worth of investing in science, and scientific leaders intensify their efforts to drum up broad support, but amidst a general sense of crisis about the apparent lack of success. This symbolic action to 'sell' science appears as just one more interest group's engagement in public relations to advance its self-interest; I will suggest later that it has deeper effects too. Having legitimated its harvest of public funds in the past by positively *distancing* itself from the ordinary public, science now finds itself politically crippled by the existence of the very cultural divide that it has helped to cultivate.

This legitimation vacuum, or public apathy, is equated by scientists with public lack of understanding of science.[3] I will argue that this is a serious misconception of the nature of public 'apathy' about science. We can say for convenience that there are two intertwined dimensions to public 'understanding' of science, the institutional and the cognitive. I will suggest that, regardless of whether there is convergence, conflict or non-contact over the cognitive aspects of science, ordinary people often have understandings of the institutional nature(s) of science in the various forms in which it impinges upon their lives. Even if they have no interest or competence in the esoteric technical discourse, they can see as well as anyone what 'social interests' lie behind a scientific position, whether the experts and their institutions are impartial and open, and they can look for social indicators of their standing and competence.[4] (As will become clear, the notion of social interests in this context is complex and requires further elaboration.) This is not to argue that they are correct in the specific judgements they make, but that this *dimension* of judgement and response is valid, reasonable and universal.

Ordinary people are logically concerned with such institutional parameters, although, as we shall see via the case-study, the cognitive elements of scientific knowledge, such as the degree of certainty or standardisation which it builds in, are also elements of a broader yet parochial cultural framework containing implicit and value-laden social prescriptions – and they are quite rationally responded to in like terms by lay people. Because in its very nature as a style of knowledge or culture, science when applied in practice often involves control and alienation, the predominant attitude towards science is logically defensive and suspicious.

If they were sovereign partners in such knowledge practices people might feel less ambivalent about this normative shift into an instrumental culture. However, because they are usually subjects for potential or actual manipulation, even when (as in for example antenatal screening, or interventions over a pollution emergency) the scientific interventions are intended to be 'for their own good', ordinary people logically feel at best ambivalent about the underlying ethos of control, fragmentation and alienation which scientific knowledge quintessentially represents. (I stress again, fragmentation and manipulation are not intrinsically negative; it depends on the context of their social practice and negotiation.) Often such interventions, even if they do not directly threaten familiar and trusted indigenous social identities, in effect do so by appearing to be incapable of recognising their moral authenticity. Thus 'science', as the language in which such interventions are justified, tends to be seen as intrinsically an adversary of such familiar social and cultural forms.

Against this general backcloth there are of course counter-examples of more constructive social relations of science. But it is difficult to make sense of public reactions to science as reactions only to its cognitive contents or immediate impacts. This basically behavioural paradigm ignores a rich and important world of symbolic relations that cannot be divorced from the empirical dimension. Indeed the empirical simply does not make sense unless the symbolic dimensions are recognised. Public responses only begin to make sense if we see them as reactions to the complete discourse of the science, in the sense of an integrated 'package' of cognitive commitments combined with social and cultural implications and interests. When this dimension is acknowledged, then the fact that science is a partial form of knowledge, concerned only with instrumental manipulation, often from within increasingly protected enclaves of power, means that there is a general background deficit of legitimation and credibility to be overcome before engagement can be established with public groups in any given case. It is important to stress that this negative account has little or nothing to do with the individual motivations and interests of scientists caught up in such processes, but has to do with the structures of ownership, control and organisation of science. The question of epistemology – what is scientific knowledge for? – is related to these social dimensions, and the whole complex forms the cultural bias of science.

In the senses outlined above, and explained via a case-study later, public 'apathy' about scientific research and development, and the legitimation vacuum within which science increasingly finds itself, are arguably due more to public understanding (of the institutional structures or 'body language' of science) than to misunderstanding (of its contents, 'rules of government' or potentials). It may be more in the domain of 'science for policy' that these deep structures of public reaction can be found, but they have practical and theoretical implications for policy for science.

PUBLIC CONCERNS – THREATS OR OPPORTUNITIES?

There is no clear-cut, objective boundary between science and the technological or other commitments associated with it. The confusion of this boundary is not a function of public ignorance, but genuine indeterminacy: where it resides at any particular time is a function of complex social negotiation and conflicting perspectives.[5] Research is sometimes justified by identifying it with a practical technological pay-off, for example embryo research and fertility treatment. On the other hand, controversial technological practices are often justified by associating them with the image of research. The nuclear industry in the 1950s and 1960s was given a strong 'scientific research front' image as part of its social legitimation; and more recently genetic fingerprinting has been presented in association with a scientific discovery image to smooth its delicate and potentially controversial uses as a social control technology.[6] In the famous case of the Harvard University recombinant DNA research moratorium imposed by the Cambridge (Massachusetts) City Council in 1976, the scientists repeatedly referred to their activities as research, whilst their opponents called it technology, with very different connotations.[7] The point is that it was both: scientists and technical experts themselves, and not just a muddle-headed public, have been actively involved in reflecting a confused and shifting picture of the boundaries of science with its surrounding technological and social commitments.

The more that science is integrated with technological and economic interests, and the more that scientists have to become promoters and 'sellers' of particular research and development programmes in a competitive market, the more deep and pervasive such real ambiguities become. In these circumstances it is difficult to dismiss as irrational or ignorant those public responses to science which ignore putative boundaries between objective, detached scientific knowledge and the associated moral, political and social connotations which are always carried with the science in particular concrete circumstances. In practice, if not in rhetoric, scientists themselves have been ignoring such supposed boundaries for a long time.

Concern about public inability or unwillingness to understand science has been based upon a false model, of scientific information floating as a kind of free intellectual good, in principle uncontaminated by any moral or social associations. Thus public ignorance, rejection or misunderstanding of science is regarded as either a fault in their constitution or a fault in the mechanisms of transfer of the information. Our findings are that people always experience scientific other knowledge as a whole package of surrounding messages and imputed interests, embodied in the *social* relationships involved in the particular communication process. This should revolutionise the way we theorise and respond practically to public concerns about science, because it demonstrates that the way science is organised, directed and controlled is an element of its public communication. In addition, it exposes the fact that people may deliberately distance themselves from opportunities for scientific understanding even when they have the intellectual capability, because of tacit moral or political evaluations of the full context or discourse in which the scientific knowledge comes to them.[8] It also means that the reasons for public adoption of science, when it occurs, may be different from those that scientists or policy-makers assume. The nature of 'public support' may also be very different from that supposed.

We have developed the analytical concept of *social identity* as the dimension shaping people's responses to scientific 'information' or other scientific interventions in their lives.[9] I illustrate the use of this concept in a case-study of the responses of Cumbrian hill sheep farmers to scientific advice after the Chernobyl accident's radioactive fall-out in 1986. I then attempt more speculatively to offer some insights concerning how better to reflect public concerns in the management of science. This broad issue was treated in the 1970s, when several bodies such as OECD funded programmes on public participation in science and technology. However, these approaches passed by a central question: what is the fundamental basis of public concern about some areas of science and technology; thus what is the impetus, or what underlies the lack of apparent impetus, for new forms of organisation, accountability or oversight of science?

There are several reasons for taking a renewed interest in the neglected question of public concerns in the management of science. One of these is an internal one. There has always been uncertainty within scientific fields as to their proper external audience, with corresponding epistemological and methodological tensions as to the rationale and principles for doing science at all. Sometimes, but not always, these tensions have revolved in extremely complex ways around grand religious and political issues, as historians of science have shown.[10] Often they have been manifested as internal debates relating to the wider social accessibility or demonstrability of scientific modes of thought and observation. There is little doubt that such tensions have been an important source of intellectual ferment and vitality within science, and it would be well for science to have some sympathetic sense of these historical currents.

A related factor in the history of science has been the critique of science as a reductionist and potentially authoritarian mode of knowledge, one which does not recognise and expressly accomodate ambiguity and negotiation. This critique has a long tradition, but its esoteric and marginalised nature has not at all meant that it was either irrelevant to science's intellectual advance or unrepresentative of a broader, more diffuse range of social concerns.[11] The critique has often resonated with one side of a recognised public ambivalence towards science and technology which is still germane today.

A third factor is that in modern times science's public role has proliferated dramatically, not only in being harnessed more and more to economic and technological innovation, but perhaps even more significant, as arbiter of disputes about the acceptability of countless new forms of technological, medical and scientific intervention in people's lives. A natural corollary of this huge expansion in science's public role has been political pressure (albeit uneven) for greater accountability and transparency in science, more stringent public concern about scientific standards, incompetence and fraud, and about bias and propaganda.[12] In other words, as its authority has grown, the basis of its authority in a vast diversity of areas has been naturally more critically examined. These shifts in the 'social contract of science' remain to be better articulated and understood, but they have been noted by several recent analysts. As one of these, Edward Yoxen, put it in a paper for the Science Policy Support Group (SPSG) in 1988, the diverse and convoluted expressions of public concern about various research developments and directions represent, not so much a threat to be contained or diverted, as an opportunity to develop the democratic foundations and resilience of future scientific research.[13] It is in this spirit that my thoughts are also offered. The essence of the route out of the legitimation vacuum is to attempt to achieve not indiscriminate 'support', but *conditional* support and to help people to develop the capability, and opportunity, to articulate their conditions. This would entail structural transformations in the social relations of science and technology.[14] Before we consider the case-study it is worth sketching some further ways in which public concerns might be regarded as relevant analytically to the direction of scientific research.

PUBLIC CONCERNS AND SCIENTIFIC DIRECTIONS: SOME ANALYTICAL BRIDGES

The profound problems created by the intensifying direct exploitation of science as a technical, economic and political decision-making resource are only just beginning to be dimly understood. As science's authority is invoked to justify more and more interventions in social lives, there is a countervailing demand for justification, whether it be by formal cross-examination of experts in the courts, public examination by critical experts, or other means. This throws into confusion the traditional notion of scientific authority generated by private specialist peer-group control. The lay public has an increasing, legitimate interest in the internal, detailed quality-control processes of many scientific research fields, and not only so as to decide which fields best deserve scarce public funds. The boundaries between 'science' and 'society' are thus more clearly exposed as social conventions, or temporary markers of a continual process of social negotiation.

There are also copious examples where lay experience and concern has indicated problems and questions previously unrecognised, but then taken up fruitfully by science. A topical example is the investigation of leukaemia clusters around nuclear installations, where firm, but imprecise public allegations about the Sellafield reprocessing plant were dismissed for nearly a decade before a television documentary forced an official investigation, chaired by Sir Douglas Black, which confirmed the clusters and began what is now a distinct scientific area involving many scientists and millions of pounds of research support.[15] Significantly, there is now a tendency for discovery of the excess leukaemias to be credited to the Black Inquiry scientists, rather than to the lay people whose knowledge caused the inquiry to happen, against the pressure of scientific dismissal of the issue.

Benguigi gives a rather different example in which a social group, French oyster fishermen, influence the development of marine biological research (aquaculture) because of their strategic position in the competition between two government agencies.[16] At the same time as the lay group played some role in defining the shape of the scientific programmes, however, the shaping scientific programmes in return influenced their practices and modes of social organisation. As Benguigi put it, the 'boundary' between science and its social environment is permeable (which is to say, it is incompletely defined and resists final definition) and the forms of interaction are not so much flows one way or the other, as (always incomplete) processes of *mutual* definition. Other have observed similar informal processes of negotiation between lay people and experts in situations where, formally speaking, there is only a one-way flow of influence and authority over the proper description of natural processes. In order to establish authority, the expert has to compromise with the laypersons' explanatory idiom(s), that is to recognise their moral standing.[17] The formal account of the interaction does not recognise this.

The need for this kind of negotiated cognitive settlement is simply that there is *always* a real question as to whether the abstract 'universal' knowledge of the expert is applicable in the particular case in hand, which is always, strictly speaking, unique: and if it can be said to apply (that is, it is thought to be better than alternatives), what conditions and qualifications attach to its applicability? The underlying processes of negotiation, of the interaction of different systems, or discourses of social identity, attempting to define the other in its own terms, but usually ending up in the compromise of *mutual* definition, were apparent in the sheep farmers and scientists case reported later.

Another form of expression of 'public concern' has emerged recently in the environmental field. Pressure groups such as Greenpeace have been growing very rapidly, with huge growth in their income. One result is that they have been able to take on more scientific research of their own, not only challenging industry or government scientific claims, but setting the scientific pace. For example, Greenpeace in 1989 appointed a 'blue chip' scientist as its Research Director, and it has injected significant new ideas and directions into, for example, the greenhouse warming scientific agenda. This is, in effect, 'public subscription' science with only a distant form of guidance over precise research directions and priorities, but with a strong mandate for the 'critical science' style, and a strong tacit statement about the lack of impartiality, quality-control and openness in established scientific institutions. This hints at the tacit institutional 'body language' point made earlier. This innovation can be seen as an attempt to re-establish a more open and pluralistic scientific ethos, akin to the Mertonian ideal,[23] that has arguably been emasculated by the institutional controls of science which have intensified over the last 30 years.

These scattered examples and illustrations do not, of course, remotely approach a systematic picture of the roles and limitations of public concern in influencing science. As a counterpoint to examples from research identifiably close to public consequences and experiences, and therefore more likely naturally to be influenced in some way, it is worth taking a field such as superconductivity. In such cases, of which there are very many in science, it is unrealistic to have public pressure or scrutiny, even of how much should be spent on it, let alone of its internal directions and commitments. Yet, even in such apparently distant fields, the connections with legitimate public concerns may be deep and potentially rapidly activated, exposing difficult dilemmas that are not resolved by the traditional resort to the language of 'scientific freedom'. For example, in the field of cytology there is an apparently purely scientific controversy over the interpretation of electron micrographs of thin biological tissues which are subjected in the course of preparation and electron microscopy to highly artificial conditions. The question is, do these abnormal conditions systematically distort what the scientists think they are seeing, when they theorise about *normal* body tissues and cell behaviour based only upon such high *abnormal* materials and observations? This is not an unusual issue within science. In this case, however, an established scientist has become increasingly uncertain and, eventually, critical of the dominant paradigm with its assumptions about the validity of the observation methods.[19] Electron microscopy is such a powerful and central technique that the very foundations of the field would be called into question were the criticisms valid. This appears to be a matter for specialists alone to resolve; but the critic has been formally isolated by institutional moves to cut off his funds and enforce early retirement, moves which have wider involvement and public reverberations. In addition, the dissenting scientist claims that it is the misconceptions created by the comprehensive dependence upon misleading methods which has obstructed scientific progress in the fields of research on degenerative diseases such as cancer.

We are now once more firmly in the public – and economic – domain, where different scientific positions reflect directly on differences over the productivity of expensive research fields supported from public money and surrounded by huge public expectations and promises. In the case of electron microscopy of cellular processes there is a strong public interest in principle (in practice it has not been expressed, so far as I know) in who is right, and what warrant scientists have for building up a colossal labyrinth of intellectual, institutional, commercial (electron microscopes are big business) and public policy commitments on the basis of assumptions which on the face of it do require more than *ad hoc* justification. the public cannot act as authoritative judge between the opposed scientific points of view, but nor can any scientist, because the scientists are implicated in the web of commitments, rather than above them. The ultimate scientific defence is that the conventional assumptions and commitments 'appear to work'; but that may itself be in dispute, both in scientific and social terms ('Are we winning the battle against cancer?').

Sociology of science helps at least put such issues in perspective as being normal to science, not pathological; it is just that most of the time they are resolved by less public social processes of negotiation, within science. The point is that in some cases they may have to be resolved, in effect, by the coarser-grained forces of public opinion (should we spend more on preventing cancer, and less on 'controlling' it?). These cases are confused by the public misunderstanding of the basic warrant for scientific knowledge and authority, which is a pragmatic and incomplete one – 'it appears to work'. This is radically different from the dominant public idea, cultivated by scientific institutions amongst others, that scientific knowledge is controlled by formal rules of method and logic, guaranteed by a professional community with boundaries distinct from politics, industry and the rest.

We should not underestimate the extent to which lay people are increasingly disempowered by the development of science and technology, as a structural matter regardless of whether funds go to Star Wars or cancer research. How far they feel the need to act against this disempowerment may, of course, depend on whether they accept the relative allocation of research funds to Star Wars or cancer research; but the backcloth of alienation is important as a fundamental basis of ambivalence and latent reaction.

In his SPSG review paper, Yoxen offered four useful categories for analysing the relationships between public concern and the steering of science.[20] These are overlapping elements of the process of 'steering': peer review and the allocation of resources; negotiation with patrons and users about middle-range goals; articulating arguments for the functions and pay-offs of research; and managing the legitimation of research, especially, but not only, negotiating around the morally contentious aspects of research. A further analytical distinction is useful between influences upon the content and directions of science on the one hand, and on its forms of organisation and regulation on the other.

It is widely agreed that there is intensifying pressure upon the effectiveness of the traditional guarantor of scientific quality and integrity, namely impartial peer review. Scientists generally have to spend more time in negotiation with external agents and interests, and the dimension of external legitimation takes on increasing significance. The central point emerges that, in negotiating to further their own field, scientists increasingly adopt the language (and role) of social engineers, because the inflating currency of persuasion involves implicit or explicit promises to give effect to some kind of technological or other social product which involves some implicit social vision.[21] Yet the inevitable corollary of this is more public anxiety, which generates even more felt need for scientists to articulate forms of legitimation in defence of a particular field, or science in general. This seems to be a self-defeating cycle of destruction of the public warrant for science in the most fundamental sense.

The foregoing sketch of the breadth of territory opened up by the title of this chapter only puts into its proper perspective the very limited empirical case which I present next, before returning to some more general reflections on what public responses towards science might mean for its management and organisation.

A CASE-STUDY: SCIENTISTS, CUMBRIAN HILL FARMERS AND THE CHERNOBYL FALL-OUT

After the Chernobyl accident's radioactive fall-out on the United Kingdom in May 1986, several upland areas of Britain suffered high levels of contamination by radioactive caesium. This caused sheep from those areas to be contaminated above the 'action level', which triggered restrictions on sale and movement of those sheep to avoid their entering the food chain. These interventions and the surrounding scientific advice caused extreme confusion and cost to the restricted farmers.[22] In Cumbria, 150 farms are still, as of 1990, restricted from free sale of sheep, and there are over 60 unresolved disputes over compensation for losses: others were settled by the Ministry of Agriculture, Fisheries and Food (MAFF), but only after bitter arguments and apparent misunderstanding of the situation by scientists and other experts. Several strands of experience and cross-cutting issues were interwoven in the responses of local hill farmers to the scientists who took up a prominent role in the management of the crisis. The situation was complicated by the proximity and the history of the Sellafield nuclear fuels reprocessing complex, as explained below. Our research interviews with local farmers, officials, scientists and others focused on the factors influencing the uptake of scientific advice by the farmers. Much of this turned upon the complex question of credibility, and this has wider social and cultural underpinnings. To explain and develop this, some background is required.

When the Chernobyl cloud passed over Britain in early May 1986, even after heavy rainfall in upland areas had washed out large deposits of radiocaesium, government scientists and ministers repeatedly reassured the public that there would be no lasting effects and that the whole affair would be over within days. It therefore came as a complete shock when on 21 June the Minister of Agriculture announced a ban on sheep sales in certain upland areas including Cumbria, where high levels of contamination had been found in sheep and lambs. Even then, however, the scientists were completely confident that this ban would be short-lived, and that the radiocaesium would soon be chemically 'locked up' in the soil, unavailable for biological uptake. The scientific model was that contamination could only come from eating the vegetation which suffered direct aerial fall-out. Once that had been eaten, the fresh growth would be clean, thus there would be no replenishment of what was a brief, one-off 'pulse' of contamination. The ban was thus set for only 21 days, and the farmers were reassured that by the time their normal annual sale time began in August (spring lambs take at least until then to fatten up on the meagre upland grass) the high levels and restrictions would be well past.

These scientific reassurances were very important for the anxious hill farmers because their heavy lamb crop in April–May, ready for

August to October sales, is their only significant source of income, and they have to be sold off the farms as soon as possible because there is inadequate grass in these meagre areas to sustain so many fattening lambs any longer. Any prospect of longer restrictions was therefore a dire threat to the whole upland sheep industry. However, after the first 21 days, against confident scientific predictions the prevailing levels did not come down, and in many cases showed increases. The ban was extended indefinitely, and new areas were included, just as the annual sales were about to start in earnest. Although the Ministry announced compensation arrangements, farmers feared complete ruin as whole breeding flocks appeared to be threatened with slaughter or starvation on utterly inadequate grazing. The restrictions involved farmers in either keeping their sheep in the hope that the radioactivity levels would fall and the free market resume, or selling them marked as contaminated and thus disallowed from slaughter but allowed to be moved to other, cleaner areas. The latter option meant big losses in the market; the former meant extra feed costs, loss of fat-lamb premiums and several other knock-on costs.

Even with this setback the scientists still believed in the essential validity of their knowledge – it was just that the biological processes must be taking a little longer than previously supposed. They therefore still advised the farmers to hang on to their sheep, even though grass was rapidly disappearing, in the expectation that they would soon to able to sell their sheep unblighted and thus at full price. Unfortunately for the scientists' credibility and the farmers' pockets and peace of mind, the scientific commitment to a particular policy stance was based upon the false belief derived from observations on lowland clay soils (and lush grass growth) that the radiocaesium would be rapidly adsorbed by alkaline minerals. This localised scientific knowledge had been uncritically extended to an entirely different context of acid organic soils and poor mixed vegetation on the fells. It was only later understood that in such conditions the radiocaesium could remain chemically mobile and biologically available for a very long time, as reflected now, four years later, in the continuing existence of the restrictions and the refusal of the scientists to make any predictions as to when they will be removed.

We can already notice one effect of the central involvement of science in the public policy decision-making process – that its process of *intellectual* commitments and their correction was tied to the timetable of a *political* crisis, and there was no opportunity even to identify let alone question the extension of scientific knowledge from one context – lowland clay soils – to another. Was this an internal scientific commitment, or an 'external' policy one? It is meaningless to try to distinguish; that is a key part of the dilemma when science is so central to policy commitments. the credibility of the science could not be separated from the credibility of the practical interventions which were based upon the science.

The practical linkage between the intellectual commitments made in science and the social, economic and other commitments made in the wider world is all-important for the theme of this chapter. The inductive extension of locally validated 'fragments' of knowledge to new conditions is normal to science. When it crashes blindly into 'mistakes' and anomalies in the laboratory, these are the meat and drink of developing scientific understanding. The problem is that more and more areas of the social and public world are being encompassed as 'the laboratory' in which science is allowed or is forced to make its mistakes. But the corollary is that, in place of a few bruised scientific egos and reputations more or less in the privacy of the scientific community, unwitting victims are created within the wider social and economic world, and they are created on the basis of commitments which are a symbiotic combination of the intellectual and the practical. Those victims, and the many more numerous onlookers, cannot possibly distinguish the science from the surrounding social commitments and processes, because at the level of commitments they are *not* distinguishable. It is logically correct, therefore, for the farmers to say, as they did, that they felt as if they were pawns in a large and uncontrolled experiment. Nor is this logic confined to them.

An important further dimension of this social basis of the science and its public responses concerns the cultural 'style' in which the science was embodied, and its incompatibility with the public culture with which it was supposed to integrate. Four elements of this cultural clash stand out – certainty and control; standardisation; fragmentation and formalisation.

Our interviews repeatedly showed a 'stylistic' clash between the farmers' modes of thought and practice and the way the science was organised and presented. This was reflected, for example, in the incompatibility of the scientific tendency to reflect certainty and control with the farmer's natural recognition (from their own practical situation) of uncertainty, lack of control and adaptation. The scientific cognitive style, or rationality, reflected traditions of control and manipulation (including its administrative–bureaucratic embodiments) which artificially emphasised certainty in public statements of existing knowledge about the contamination. The farmers' rationality reflected a long-standing tradition of continual adaptation to a range of key factors on which they depended, yet over which they could not exert control, such as climate, markets, government policies, National Park Authority and landlord interventions, and so on.

In similar fashion, the farmers were well attuned to variability – of local environments, sheep habits and optimal farm management differences even within the same valley. They thus received with profound scepticism the *standardised* format of scientific statements (independent of what, explicitly, such statements said about the contamination). Again, these cognitive incompatibilities were rooted in social and cultural differences between the farmers and scientists.

Thirdly, it was clear from farmers' responses to scientific advice over Chernobyl that they evaluated that advice in more comprehensive terms – of the credibility of the institutions managing the whole situation. For the farmers, those institutions were continuous with those responsible for controlling the Sellafield nuclear plant, which has been locally (and internationally) controversial for many years. In particular the experience of many farmers of the secrecy surrounding the 1957 fire, which spilled tens of thousands of curies of radioactive contamination onto the same area, deeply affected their view of the Chernobyl situation. Many expressed the belief that at least some of the contamination now being highlighted was from Sellafield, but had been covered up or studiously ignored. In other words, the real embodiment of 'scientific' statements about the 'Chernobyl' contamination was the whole sorry institutional track records of what they saw as official secrecy, lies and misinformation about their local Sellafield plant. What is more, from their position this more historical, relational framework was utterly rational as the framework of response (whatever its particular content – it could conceivably have meant greater trust and credibility over Chernobyl had there been a different history).

These clashes coincided with a similar cultural dislocation over how to respond to the new situation. The administration entered with elaborate documents (which few farmers ever read), formal requirements (for example to give several days' notice of intended sheep sales, so that monitoring could be organised) and rules which simply did not recognise the informal, flexible *ad hoc* and adaptive style of hill farm decision-making. At one angry meeting between farmers and MAFF experts, nearly a year into the crisis, an expert told the farmers they should have been keeping a full, detailed record of all extra costs associated with the restrictions, from day one. Such a bureaucratic method was completely alien to the farmers, who relied upon personally-known Farmers' Union officials to help them with their tax forms and compensation claims; and, as one irate farmer pointed out, in the early days the same experts had been telling them the restrictions would not last, so why should they have kept any records anyway?

These *cultural* dimensions of the lay response to science-as-a-culture can be drawn together into an explanation of the farmers' overall response to science: it was a response fundamentally rooted in the concern to maintain a social–cultural identity, in the face of threats represented by more general interests and trends with which science was seen to be associated. Likewise, scientific responses to scepticism, opposition or neglect can be seen as essentially symmetrical, as social responses focused on maintaining a particular socio-cultural identity, one which happens usually to be the more powerful institutionally. For example, our interviews often revealed contradictory models of science being expressed by the same respondent over the course of a single interview (usually one to two hours' length). The context of each model's expression is important. In one, a model of 'science as omniscient' was expressed. In this model, scientists knew all along that the contamination would last and remain available a long time, but entered into a conspiracy with the authorities to deceive the farmers and get them to hang on to their sheep rather than sell them short and receive compensation. In the other model, the scientific knowledge is always incomplete and uncertain, but scientists are too arrogant to recognise that, or to recognise other forms of knowledge needed to relate their own theoretical understandings to reality. Thus the farmers gave copious examples of the scientists' ignorance and neglect of specialist farming knowledge and expertise, and bemoaned the neglect of this less formalised, but no less expert nor less necessary knowledge. This specialist hill farming expertise was part and parcel of their social and cultural identity, and they felt this to be threatened by the official lack of interest in their knowledge. This was exacerbated by the official demand for formal methods, accounting (for compensation), decisions and planning, and so on, in a way that was not only alien to, but impracticable for, the farmers.

These contradictory models of science – 'science is omniscient' and 'science is incomplete' – can be understood as consistent once a different level of analysis is adopted, that of lay people defending a threatened social identity. Broadening the focus of the Chernobyl issue, hill farming in National Parks like the Lake District has been under intense threat for several years. It is always the most marginal sector of farming, and farming generally has felt left out by its traditional ally, a Conservative government. It has also been under great pressure in the Lake District National Park from tourism and from management of the hill farming area by people felt to be unsympathetic and ignorant toward hill sheep farming. Thus the farmers have been threatened by suggestions that they should in effect become merely park wardens, managing sheep only to beautify the landscape for tourists, not to make an economic livelihood. Add to this pressure on housing from holiday home 'offcomers', leaving some villages up to 50 per cent empty in winter, and of farm landlords such as the National Trust who are often felt to be less than sympathetic to farming needs, and there is a considerable sense of crisis over the very existence and survival of the traditional hill farming society. This background of external pressures and internal insecurities is the context of the Chernobyl crisis's arrival. It is not surprising if farmers felt neglected, excluded from managing their own affairs, threatened and even conspired against by officialdom, including science.

That people hold contradictory models of science could be taken as a sign of their intellectual weakness on the conventional approach, which is based on a 'scientific' framework. However here one can see a deeper rationality where the 'omniscience' model is deployed to make sense of one set of significant experiences and the 'incomplete yet arrogant' model makes sense of other significant experiences in the same domain. In the former their identity was threatened by conspiracy to undermine them economically; in the latter their social identity was threatened by neglect of their own expert specialist knowledge of hill farming – an expertise which was a central element of their identity.

This interpretation of the basic dimensions of public response to 'science' is consistent with the more detailed cultural analysis of the nature of cognitive dislocations and disagreements between farmers and scientists. Although in the case of the Cumbrian hill farmers the notion of social identity may have a more recognisable form than in many others, I suggest that it offers a generally more authentic and productive means of understanding interactions between science, technology and lay publics than existing frameworks based upon the central idea that people respond to violations or confirmations of 'nature', or of their cherished 'values' or 'interests'. 'Identity' connotes something deeper, more relational, less complete and more involuntary than mere 'interests', or even 'nature'.

Thus to summarise:

1. Surveying people about their 'attitudes to science' is more or less completely useless, because it wrongly presumes control of an uncontrolled variable – what people mean by 'science'.[23] As our research has shown, it can mean different, apparently contradictory things even to the same people, depending upon its subordinate place in more fundamental constructs.
2. People's concerns and the factors informing them are not primarily about nature and its violation or confirmation, but about fundamental social identities and relationships. 'Nature' and 'natural knowledge' are contingent and negotiable elements of rhetoric in the endless processes of building, maintaining and developing such social identities.
3. The ethos of science is fundamentally about control, standardisation and manipulation. This is not only an intellectual temperament but a material, social and cultural form. Whatever the cognitive content of particular items of scientific and lay beliefs, many areas of social life do not recognise an ethos of control and manipulation as meaningful, therefore in those domains they cannot even begin to give credence to cognitive frameworks like science which speak in such terms (and which in effect, prescribe such relationships). There is an inalienable moral and social dimension to 'cognitive' interaction between science and lay beliefs. The sheep farmers could not respond to the scientific advice on the Chernobyl fall-out without seeing its (to them) inappropriately certain, standardised and formalised properties, which were prescriptive as well as descriptive. Nor could they create an artificial distinction between the institutions controlling the Chernobyl emergency and those controlling Sellafield's history. They responded, inevitably and reasonably, to this whole culture of 'knowledge, institutions and social values' together.
4. Thus policy and research concerns about public uptake (or lack of uptake) of science must focus on the underlying cultural elements of social organisation, control and relationships implied by scientific frameworks in a given field. Social uptake may be

improved more by changing the organisation and culture of science than by repackaging the scientific message or redesigning the intermediate transmission process(es). Indeed it must be recognised that the concern about lack of social uptake of (i.e. support for) science expressed by scientists themselves is a reflection of the structural reality in which science is controlled, organised and applied in such a way as to expropriate, reduce and otherwise threaten people's familiar social identities, rather than to recognise and negotiate with them.

Although it seems obvious to say that public concerns expressed strongly in relation to certain areas of science, such as embryo research and genetic manipulation, are about violated social and personal identities rather than cherished ideas of nature, our research has shown that this is true more generally for areas of science such as radioactivity that, on the face of it, have little to do with transforming social relations or human constitutions.

CONCLUSIONS

I have focussed mainly on a case-study in the domain of science for policy, while the usual focus in the management of science is policy for science – what areas of research should be funded, via what institutional mechanisms? The linkages between these two conventionally distinct domains in science policy studies have been left undeveloped and under-investigated, for one reason because there has indeed been a large social–political gap between them. About policy for science there has been resounding public apathy in general, (though on specific issues this has not been true, for example the call for R&D in renewable energy) whilst in science for policy the public reactions usually studied can most frequently be characterised by a political reflex of self-defence against science seen to be too often ranged against communitarian concerns and in favour of big economic and technological interests.[24]

The analysis I have offered shows why the former, and the legitimation and public support crisis it now creates for science, can be seen to be a direct relation of the latter. However our research has revealed a depth to public suspicion or neglect of science which goes well beyond the analytical treatment typical of the 'science for policy' field. In this research, public responses tend to be conceived in terms of immediate interests – threatened health, livelihood, amenity etc.[25]

We have found that, even when there is no such threat or confrontation apparent, people may simply divorce themselves from science because they construct their social identity in such a way that 'science' is irrelevant, or associated with more indirect and diffuse, but no less real, threats and antipathies. In the Chernobyl case there was no *a priori* reason for the scientists and lay people (the hill farmers) to be in confrontation. This emerged from deeper than mere 'interests', and from somewhere different from objective threats to physical environment and health. I have suggested that this source of non-communication – the basis of the public legitimation vacuum–lies in the dimension of social – cultural identity.

The orthodox view is that people respond to scientific knowledge, rather than – more realistically – to the total social complex of which it is a part in any given case. Corresponding to this is the view that expressed concerns about technological risks and environmental threats – the major contemporary arena in which conflicts and confusions exist about public assimilation of scientific expertise – are founded in perceptions that physical safety, or nature, has been violated beyond some acceptable threshold. We argue, contrary to this, that the more fundamental category influencing such reactions is not physical nature but social identities. This does not mean, of course, that categories of 'nature' do not play a role in such social and cultural constructions. However the key point is that our suggested framework offers an explanation as to why natural scientific knowledge has intrinsically limited purchase on public attitudes about, for example, the environment (for other reasons than just that they 'cannot understand'); it reveals that the language of physical environments, over which we might expect science to have authority, is only a surrogate for deeper and more complex *social* anxieties, for which, as the case-study illustrated, science is, often unwittingly, more part of the problem than part of the solution.

Thus we find an explanation of the common policy experience of intensifying the *scientific* framing of such issues, only to find growth, not reduction, of public concern and criticism, because their more authentic concerns are being further denied by the scientific idiom which in effect claims that only physically-based meanings exist. This leads directly to the central idea of this chapter: the suggestion that, in its present mode, science is in what can be seen as a negative, if slowly unfolding, spiral of the self-destruction of its own public credibility, or automatic de-legitimation. The underlying general structure of this process appears to be the following:

1. Scientific research increasingly adopts the language of social engineering, if for no other reason than that it is increasingly forced to 'sell' its programmes (and possible products) in order to survive. In the environmental policy field, as we saw, it is pressed into direct service in many practical situations where the parameters of the practical context have not been properly researched, so that the fit of pre-existing scientific understanding to the new situation is a matter of (often highly exposed) experiment. However it is experiment without the principal actors even realising it, and with already-committed material decisions and public claims – that is, with the credibility of political and scientific institutions – hanging in the balance;[26]
2. Either way people experience science as a normally external force associated with social programmes, future visions, implicit prescriptive models of social relationships, that may well be deeply threatening to their own familiar social–cultural identity, all the more so if these models are not even being acknowledged by their bearers;
3. Not recognising this constitutive social architecture in which science is experienced, scientists and policy elites, in the face of public scepticism or worse, persist in amplifying their scientific framework of 'control', thereby further denying the legitimacy and authenticity of the social concerns, thus provoking further social incredulity and hostility – and so on in a vicious spiral.

This argument does not at all deny that there are many and diverse countervailing experiences. Furthermore, although the instrumentalism of scientific knowledge always contains the potential for conflict with communitarian experiences and concerns, there is nothing inherent or inevitable to this, because all social identities must as a matter of necessity sometimes live with, indeed require, manipulation and instrumentality. Thus the source of problems must be sought in the structure of control and coupling between science and society. Arising out of *structural* features of the social organisation and control of science, there is a deep ambivalence which offers a threat to the public warrant for science itself, in the most fundamental sense. Instead of assuming that public lack of understanding of or interest in science is a problem in either the public's mind or in the techniques of 'communication', further research, and policy, should face up to this dimension also. In addition to copious empirical evidence that people do not like what they see as the 'capture' of science by particular interests, one could read widespread public disaffection with established scientific institutions into the mushrooming public subscription to environmental groups like Greenpeace which are increasingly spending that money on campaigning via critical science, which at its best opens up established science to more invigorating and searching debate. Indeed this is tantamount to an attempt to re-establish the ideal of science as an open, pluralistic democratic culture, restricting rather than reinforcing the unhealthy concentration of economic or political power.

The reader coming from conventional science policy may recognise resonances between the present points and the emphasis placed in recent years in organising 'upstream' R&D, so as better to accommodate inputs from potential users about the users' *real* situtation, as opposed to often naive models of that situation held by the laboratory researcher.[27] The now official Great British Fiasco of nuclear power could be attributed to a similar syndrome in which academic scientists in the institutional guise of the Atomic Energy Authority dominated policy choices for so long in the early decades and repeatedly made mammoth technological and surrounding policy and social commitments without realistically assessing the more mundane 'downstream' engineering and operational questions.[28] (As an aside, it would be worth investigating comparatively whether this was a peculiarly British syndrome, brought about by the extravagantly *scientific* idiom in which the British programme was legitimated to the public at large.)[29]

This practical point about openness of research to 'downstream' social actors reinforces the theoretical perception of scientific knowledge as *inherently* incomplete, in that when it is applied to new circumstances a model of that new situation has to be deployed, and this involves selecting the significant and insignificant features of that situation. Especially if these situations involve other people not participating in the research process, they may have legitimate experience and evaluative stances about these matters. Thus negotiation with application-context people or users is essential to *complete* the scientific knowledge. Of course scientific knowledge has to be allowed to idealise and simplify some of its context in order to proceed ('let us assume a perfectly elastic solid ...'). However the more that it is involved in application the more that lack of awareness of these artificial commitments is dangerous and potentially provocative.

I would assert that management of science has not yet taken this point, with its implications about the need to arrange for more open political and social influences on what are usually unreflective and tacit social models informing laboratory research. This does not mean that laboratory research should become political seminar, but it means deliberate defence of organisational diversity and cross-criticism in research. Indeed it is difficult to see how this 'downstream responsiveness' issue could be addressed without, *inter alia* a deliberate attempt to counteract the tendency towards private control of research, because it would entail a wider degree of participation in technical negotiations normally conducted only between scientists without exposure of their assumptions or practices beyond a very confined organisational circle. Correspondingly, the message is that the trend towards privatisation of research probably means *less* ability to incorporate user and application experience in the formation of knowledge. This in turn must entail large costs against this institutional mode of science.[30]

The familiar question of social interests concealed in scientific knowledge emerged earlier. This is not the place for an extended discussion of what has been a well-worked topic.[31] Coinciding with part of the social interests debate, I do not think the term 'social interests' adequately describes the panoply of structural elements, of a whole social, administrative, cognitive and cultural discourse which I am suggesting lay people experience in encounters with 'science'. This partly relates to the problematic nature of the concept of social–cultural identity which I introduced as a favoured explanatory approach. This concept is itself not just fuzzy but essentially incomplete, awaiting practical (but temporary) completion *through* social interaction. Thus the imputation of 'interests' concealed in scientific knowledge – as in the sheep farmers' inference of a political conspiracy to undermine their livelihood (and thus identity) in the scientific statements about the short life of radioactive contamination – is structured itself by the crystallisation of social identity: in this case, 'here *we* are, hill farmers, under threat from the conspiratorially-shaped science' (to paraphrase).

At least partly, therefore, imputed 'interests' in scientific knowledge which people experience, and their own fundamental social identity, both help to define each other, in this case (but not always) in contradistinction, in the material process of interaction. The hill farmers' imputation of conspiratorial social interests in the science may in this case have been incorrect. However it, too, was an oblique way of expressing a more complex truth, about the utterly alien and incompatible culture forced upon them by the post-Chernobyl emergency.

Finally, I raise a question about what might be intrinsic structural limitations on public interest in science, since scientific and policy concern about a 'legitimation vacuum' is where we started. It is worth at least considering that limited public interest in and support for science is simply a natural function of lack of power, a belief that they cannot exercise any power to control science, so why make the effort to understand (and anyway risk social degradation if one does not fully succeed)? This view would be based upon a social model of latent conflict and alienation, in which people feel trapped by power, going along with it *faute de mieux*, but with little or no sense of identity with it and its legitimatory culture of science, technology and so on. (This does not mean, of course, that they will not take benefits where they can, opportunistically.) In the typical perspective of experts and policy elites, lack of public interest cannot be due to fundamental social alienation because they take for granted an integrated, consensus model of society. Thus the only alternative form of explanation of the intractable lack of public interest is that the public must be uneducated about the matter.

On the other social model, the problem of public understanding of science is intrinsic to structures of power in society. We can, and should, work on it at its margins, but we should not pretend we are engaged in anything more fundamental. It is tempting to suggest that the public understanding of science problem has emerged, not for anything the public has done or not done, but because scientists have become insecure about the deepening bite of the structural process of de-legitimation explained above. Seeking to restore a political mandate for science, they have looked to educate the public, when they could perhaps still spend some time educating themselves. In addition to transforming public cognitions – indeed *in order to* transform them – they will have to work to transform the institutional structures of science, and thus their own social identities.

NOTES AND REFERENCES

1. See, for example, J. Brown (ed.), *Environmental Threats: analysis, perception, management* (London: Belhaven, 1989).

2. For example, in the area of embryo research and genetic manipulation. See 'Scientists warn of dangers of mob rule', *Observer*, April 1989. For an argument in favour of a new social contract of science which involves extended peer-group communities, very much in parallel with the present analysis, see S. Funtowicz and J. Ravetz, *Global Environmental Issues and Post-normal Science'* (London: Council for Science and Society, 1990).
3. The Royal Society of London, *The Public Understanding of Science* (London, 1985).
4. B. Wynne, 'Technology, risk and participation: on the social treatment of uncertainty', J. Conrad (ed.), *Society, Risk and Technology* (London: Academic Press, 1980) pp.83–107.
5. For an insightful discussion in the field of risk assessment, see Sheila Jasanoff, 'Contested boundaries in policy-relevant science', *Social Studies of Science*, 17, (1987) pp.195–230; also T. Gieryn, 'Boundary-work and the demarcation of science from non-science', *American Sociological Review*, 48 (1983) pp.781–95.
6. G. Myers, 'Unmistakeable Identity: actors and events in news reporting of DNA fingerprinting', Lancaster University, mimeo, 1989.
7. S. Krimsky, 'The recombinant-DNA dispute', in D. Nelkin (ed.), *Controversy* (London/Beverly Hills: Sage, 1979).
8. These could be personal and historical (they have been put down by their science teacher at school) and/or more political (they believe that the scientific knowledge is being or will be put to unacceptable use, and do not want to be identified with, or ensnared by, the whole social–technological–moral direction of which the scientific knowledge is seen to be an element).
9. As indicated later, and discussed at greater length in other papers from the Lancaster project, the concept of social or cultural identity is not at all unproblematic. But it does at least move the issues on to a better plane.
10. For an exemplary study of this kind, see S. Schaffer and S. Shapin, *Leviathan and the Air pump: Hobbes, Boyle, and the Experimental Life* (Princeton: Princeton University Press, 1985).
11. Again, there is a huge literature on this topic, but see for example, M. Berman, *The Re-enchantment of the World* (New York/London: Bantam Books, 1984).
12. D. Nelkin (ed.) *Controversy* (London/Beverly Hills: Sage, 1979).
13. E. Yoxen, 'Public Concerns and the Steering of Science', London, Science Policy Support Group, March 1988.
14. D. Layton, 'Transforming science for public access', University of Leeds, Dept. of Education, mimeo, 1986.
15. For an indication, see the annual Reports (First Report, 1986) of the Government Committee on Medical Aspects of Radiation in the Environment (COMARE) set up as a Recommendation of the Black Inquiry.
16. G. Benguigi 'The Scientist, the fishermen and the oyster farmer', in S. Blume (ed.), *The Social Direction of the Public Sciences,* Sociology of the Sciences Yearbook (Dordrecht:Reidel) (1987) pp.117–134.

17. See, for example, C. Helman, 'Feed a cold, starve a fever', *Culture, Medicine and Psychiatry*, 2 (1978) pp.107–137.
18. The Mertonian model of science as an ideal democratic republic of co-operation, complete egalitarian sharing of ideas, impartial critical examination of knowledge claims and meritocratic distribution of recognition and status, has been recognised as a 'functional myth', rather than a description of scientific practice. See, for example, S.B. Barnes and D.O. Edge (eds), *Science in Context* (Oxford: Open University Press, 1982) especially pts 1 and 2.
19. H. Hillman, at the University of Surrey. As I hope the discussion makes clear, I am in no position to judge which side is correct, and my analysis does not require such a judgement.
20. Yoxen, 'Public Concerns', 13.
21. For a systematic and insightful, though ultimately, I suggest, flawed analysis of these dimensions, see B. Latour, *Science in Action* (Milton Keynes: Open University Press, 1987). One of the questionable aspects of the approach of Latour and Callon is their conflation of analysis and engineering. Thus they refer to scientists as expert social analysts in plotting the reorganisation of society so as to enlarge their scientific or technical 'empires'. Whilst there is a warrant in instrumentalist epistemology for equating social engineering with social analysis, there are many cases where scientists inadvertently employ naive social models and assumptions about contexts of application, where often elementary social analysis could have adapted their understanding and their strategies. The Latour–Callon approach implies a degree of deliberate, conscious intent to manipulate the social world which may be *post hoc* gratuitously reading singular volition into a more multivalent, less determind social process of interaction across several social frameworks and localised interests. To reduce all analytical possibilities to instrumental manipulative intervention is to imply that such intervention is always 'strategic', in the sense of planned against a systematic method of feedback, correction of original assumptions and revision of strategies.
22. This case has been partly written up in B. Wynne, 'Sheepfarming after Chernobyl: a case-study in the public communication of science', *Environment*, March 1989, pp.11–17, 31–39.
23. This fallacy seems to have been perpetuated in the large-scale surveys of public attitudes to science. For an example, see J.R. Durant, *et al.*, 'The Public Understanding of Science', *Nature*, 340, 6 July 1989, pp.11–14. Distinction must be made between attitudes to, and understanding. If one asks people in our society in such surveys whether the earth goes round the sun, or vice versa, one can be reasonably assured that there is a commonly shared meaning of the terms 'earth', 'sun' and 'goes round' across all the respondents, and between them and the researcher. Their answers can be gauged against established scientific understanding. But if one asks people whether they believe 'scientists can be trusted to make the right decisions' (as was asked in the Durant *et al.* survey) then uncontrolled variation is almost certain to

enter into the responses as to what people mean by 'scientist', 'right decisions' and 'trusted'.

24. Nelkin (ed.), *Controversy*, op. cit., 12.
25. Ibid.
26. See, for example, B. Wynne, 'Unruly technology: practical rules, impractical discourses and public understanding', *Social Studies of Science*, 18 (1988) pp.147–67; W. Krohn and P. Weingart, 'Nuclear power as social experiment: political fallout from the Chernobyl nuclear meltdown', University of Bielefeld, mimeo, 1986. Also, Krohn and J. Weyer, "Society as Laboratory" Bielefeld, mimeo, 1989.
27. See, for example, E. von Hippel, *The Sources of Innovation* (Oxford University Press, 1988); C. Freeman, *The Economics of Industrial Innovation* (London: Pinter, 1982).
28. D. Burn, *The political economy of nuclear power* (London: Institute for Economic Affairs, 1967) contains this kind of critique of the UK programme, but with a strongly free market promotional slant.
29. I laid this as a suggestion in B. Wynne, *Rationality and Ritual: The Windscale Inquiry and Nuclear Decisions in Britain* (Chalfont St Giles, Bucks: British Society for the History of Science, 1982) but never followed it up.
30. I hesitate to call this shift a reflection of 'science policy'. It is simply the current political economy of science.
31. For a useful case-study and discussion of the idea of cultural identity as a determinant of intellectual constructions, see G. Downey, 'Reproducing cultural identity in negotiating nuclear power: The Union of Concerned Scientists and Emergency Core Cooling', *Social Studies of Science*, 18 (1988) pp.231–64. For the 'interests' debate, see S. Shapin 'Following Scientists Around', *Social Studies of Science*, 18 (1988) pp.533–50; S. Woolgar 'Interests and Explanation in the Social Study of Science', *Social Studies of Science*, 11 (1981) pp.365–94. See also Correspondence in *Social Studies of Science*, 11 (1981) pp.481–514.

Index

academia and industry links xii, 58–9, 94–5, 127, 139
in biotechnology 116, 119–20, 123–6
accelerators 52, 99, 100, 101
accountability principle 1–2, 57, 65
Accounting Standards Committee 67
Addison, Christopher 53
Advisory Board for the Research Councils (ABRC) 2–19 *passim*, 55, 64, 89
see also Council for Scientific Policy
Advisory Council for Applied Research and Development (ACARD) 55
Advisory Council for Science and Technology (ACOST) 55, 65
aerospace industry 133–40, 142
Agricultural Genetics Company (AGC), UK, 120
Agricultural Research Council (ARC) 53
agriculture and biotechnology 113–14, 123–5
Agriculture and Food Research Council 2, 7
Akzo 124
allocation of resources
in science 1–2, 5–11 *passim*, 13–15, 83; public perceptions 152
Alvey Information Technology Programme 55, 93–4
AmGen, USA 117
applied research *see* research, tactical
apprenticeship schemes, in technology 62
'arms-length' principle, government and research 53
Arrow, K. 27
Ashai Chem, Japan 124
Ash, Sir Eric xi
astronomy 66
Atomic Energy Authority 164
Atomic Energy Research Establishment 99
Australia 53–4
Averch, H. 27

Bannier, CERN 104–7
Bartlett, Sir Frederick 50
BASF 124
basic research *see* research, basic
Bayer 124, 127
Behrman, J. 138–40
Belgium
and CERN 104
research expenditure 23–4
technology 36–7
benefits of research, evaluating 10–13, 57–58, 87–8, 140
Benguigi, G. 150
bibliometrics 38, 42–3

biochemistry 113
biology, citation indices 60
biotechnology xi–xiii, 32, 40, 50,112–29
academia and industry links 116, 119–20, 123–6
defined 112–13
interdisciplinary research 115, 126
and small firms (NBFs), 115–23
and venture capital 116–19
Biotechnology Investments Ltd (BIL), UK 118
Black, Sir Douglas, and the Black Inquiry 149
Blue Streak rocket 106
Board of Longitude 51
Boulton, Matthew 58
Boyle, Robert 48
'brain drain' 139–40
Branscombe, Lew 67
British Association for the Advancement of Science ix
British Biotechnology 117
'British Empire Syndrome' 3, 9
Bull, E.T. 113

Calgene, USA 117
Canada, technology 36–7
Carpenter, M. 30
cell fusion 113
Celltech, UK 117, 120
Central Council for Science and Technology 55
Central Policy Review Staff 55
Centre National de Recherche Scientifique (CNRS), France 120
'centres of excellence' 65
CERN 3, 9, 98–111
Cetus, USA 117
chemical industry 30, 59, 133–40, 142
chemistry, citation indices 60
Chernobyl 147, 153–61, 162, 165–6
Chiron 124
Ciba-Geigy 123, 124
citation indices 31–3, 60, 122
clinical medicine, citation indices 60
Clurman, Harold 87
Cockcroft, Sir John 99
cold fusion 76
Coleman, R.F. 117
collaborative research 65–6, 82, 93–5
Committee on Earth Sciences (CES), USA 77–8
Committee on Global Change, National Research Council, USA 77
computer industry 133–40, 142
Corning Glass 124
corporate planning x–xi, 5–7, 57, 65
costs, research xi, 70–2, 89–90, 140–1
Council for Science and Society 35
Council for Scientific Policy, UK 53, 60
Cumbrian sheepfarmers *see* Chernobyl
Curie, Marie 50
cytology 151–2

Dainton, Lord x–xi
Darby, Abraham 58
Darwin, Erasmus 58
Dasgupta, P. 27, 41
David, P. 29
de Rose, Comte François 104, 106–9
de Tocqueville, Alexis 26
Denmark 127
Department of Agriculture, USA 77
Department of Defence, UK 52
Department of Education and Science (DES), UK 2, 53, 55, 63–4, 90
dual support system 55, 56, 64, 66, 67
Department of Energy, USA 77
Department of the Environment, UK 52
Department of Health, UK 52

Department of the Interior, USA 77
Department of Scientific and Industrial Research (DSIR), UK 53, 99, 102, 107
Department of Trade and Industry (DTI), UK 94–5, 127
DES *see* Department of Education and Science
developmental research *see* research, developmental
Dicken, P. 139
Directory of British Biotechnology 116–17
division of labour 3–4
DNA 113, 146
Dosi, G. 113
Dow Chemical 124
DSIR *see* Department of Scientific and Industrial Research
Du Pont 124
dual support system of funding 55, 56, 64, 66, 67

Eads, G. 28
Economic and Social Research Council, UK 2, 5–7, 11–13
economics, and management of science 1–17, 26
economics, management, science relationship 9–17
education
  in science and technology 22, 25, 26, 32–4, 41–2
  see also skills
efficiency evaluation, and science policy 57
ELDO 106, 109
electromagnetic induction 86
electron microscopy 87, 151–2
electronics industry 30, 32, 133–40, 142
embryo research 146, 161
energy, renewable 161
engineering, citation indices 60
engineering industry 40
Enichem 124
Environmental Protection Agency (EPA), USA 77
environmental research 71, 77–9, 150, 164
ESPRIT, EC-funded technology scheme 132
ESRC *see* Economic and Social Research Council, UK
ESRO 106, 109
European Community (EC) 127, 132
European Nuclear Research Centre *see* CERN
European Space Agency 55
expenditure on research 9, 13–14, 60, 62, 121–3
  evaluation 5–6, 10–13, 57–8, 63–4, 87–8, 140

Fagerberg, J. 25
Faraday, Michael 58, 86
Federal Co-ordinating Council for Science, Engineering and Technology, USA 77–8
fermentation technology 113
Fischer, W. 138–40
Flowers, Lord 91
Frame, J. 31, 32, 38
France
  and biotechnology 116–17, 119
  Centre National de Recherche Scientifique, 120
  and CERN 98, 104, 105–9
  citation indices 60, 122–3
  Institut National de Recherche Agronomique 120
  Institut Nationale de Santé et Recherche Médicale 120
  Institut Pasteur 120
  research expenditure 23–5, 108–9, 121–2, 134–5, 137
  technology 32, 36–7, 61
'free good' concept of basic research 28–9, 35, 40
freedom, intellectual 34–5, 54–56, 65, 78, 95–7
Freeman, C. 30
funding for research 26–30, 31, 41, 89–95

funding for research *(cont'd)*
  in CERN 98–109
  from overseas xii, 35–7, 40, 41, 131–42
  in new technologies 112–28

Gabor, Denis 87
Genentech, USA 115, 117, 124
genetic fingerprinting, social implications 146
genetics and genetic engineering 113, 146, 161
Germany, West
  and biotechnology 116–17, 119–20,
  citation indices 60, 122–2
  development of technical skills 39, 62
  Gesellschaft für Biotechnologie Forschung 119
  research: expenditure 22–4, 27, 121–2, 134–5; links with industry 59, 123–5; restrictions 127
  technology 32, 36–7, 61
Gesellschaft für Biotechnologie Forschun (GBF), W. Germany 119
Gibbons, M 125
Glaxo 125
government and science
  relationship: changing 62, 64–7, 74–5; economic 26–30, 40–2; evolution 52–62; post-war 55–8, 59–62, 73–4, 84; UK 17–20, 51–67; USA 69, 83
  skills provision: UK 41; USA 72, 81–3
  UK: 'arm's-length' principle 53; role of civil servants 17–20
Gramm, Rudman, Hollings Act, USA 73
'green' movement xii, 14, 150
'greenhouse warming' 150
Greenpeace 150, 164

Haldane, Richard Burdon 53, 59
Hare, P. 38
Harvard University, USA 146
Harvey-Jones, Sir John 96
Health and Safety Executive, UK 52
Heyworth Report 53
Hicks, D. 38
Higgs, Dr Peter, and the 'Boson' 14–15
high-energy physics research 98–111
hillfarmers *see* Chernobyl
Hoechst 124–5, 127
holography 87
Holt, G. 113
Hood, N. 138
Humboldt, Karl Willhelm von 59
hydrogen bubble chamber 101

IBM 67
ICI 96, 124, 125
Illinois Institute of Technology, USA 34
Imperial Chemical Industries *see* ICI
Imperial College of Science and Technology, UK 89, 90, 91
Industrial Revolution, UK 58–9
industrialisation, role of science 58–62
industry and academia links xii, 58–9, 94–5, 127, 139
  biotechnology 116, 119–20, 123–6
information dissemination 30–4, 35–7, 42, 59, 82
innovation, technological 34–5, 61, 62
Institut National de Recherche Agronomique (INRA), France 120
Institut National de Santé et Recherche Médicale (INSERM), France 120
Institut Pasteur, France 120
Institut for Biotechnology, W. Germany 124
instrumentation costs of research xi, 71–6 *passim*
insulin 113

intellectual freedom 34–5, 54–56, 65, 78, 95–7
intellectual property rights and patents 91–3, 114
interdisciplinary research 65, 77–8, 81, 115, 126
interferon 113
international research organisations xi, 3, 66, 138
  CERN 98–111
  ELDO 106, 109
  ESRO 106, 109
  European Space Agency 55
investment in research 27–30, 40, 67, 72, 116–19
Irvine, John 22, 25, 31, 33, 121
Isard, P. 25
Italy
  and CERN 98
  research expenditure 22–4, 134–5, 137
  technology 36

Japan
  citation indices 60, 122–3
  development of technical skills 39
  research expenditure 22–5, 121–2, 134–5
  technology 32, 36–7, 61
Jewkes, J. 28
Johnstone, R. 31

Kay, J. 27, 35
Kodama, F. 22
Krige, John xi, 9

Laboratory of Applied Psychology, Cambridge, UK 50
Laboratory of Molecular Biology (LMB), UK 119
lasers 42
leukemia and nuclear industry 149
Levin, R. 30, 33, 34
Liebig, Justus von 59
Lilly, M. 113
LINK scheme, UK 94
Llewellyn Smith, C. 27, 35
Lockspeiser, Sir Ben 99
Lunar Society, USA 58
Luukkonen-Gronow, T. 43

McAllister, P. 38
MAFF *see* Ministry of Agriculture, Fisheries and Food
magnetron valve 59
management, science, economics relationship 9–17
Manchester Business School, UK 12
Manhattan Project 73
Mansfield, E. 29, 34
marginal cost analysis, and science 12
Martin, Ben 25, 31, 33, 121
Marx, Karl 26
Massachusetts General Hospital, USA 125
Massachusetts Institute of Technology, USA 32, 89
Max Planck Institutes, W. Germany 119, 120, 124
measurement techniques 42–3
Medical Research Council (MRC), UK 2, 5, 7, 49–50, 53, 119
medicine, clinical, citation indices 60
Melville, Sir Harry 107–8
meta-systems, and management of science 9–17
meteorological research 71
Middleton, Sir Peter 1, 2
Miescher Research Institute, Switzerland 123
Miles Cutter Laboratories, USA 124
Miller, Roberta B. xi
Ministry of Agriculture, Fisheries and Food (MAFF), UK 154, 157–8
  and research, 52
Ministry of Defence, UK 90
Ministry of Science, UK, proposed 53–4
Mitchell, Professor E.W. 99
Mitsubishi Chem 124

Mobay Research Centre, USA 124
molecular biology 50, 71, 113
Monsanto 124
Montedison 124
Morgan, Gareth 13
Morris Report 65
Mowery, D. 25, 27
MRC *see* Medical Research Council
multidisciplinary research *see* interdisciplinary research
multinational companies 35–7, 41–2, 132, 138–9, 142

Narin, F. 30, 32, 34, 38
National Aeronautics and Space Administration (NASA), USA 77
National Environmental Research Council, UK 53
National Institute for Research in Nuclear Science (NIRNS), UK 100
National Oceanic and Atmospheric Administration (NOAA), USA 77
National Parks 159
National Radiological Protection Board, UK 52
National Research Council Committee on Global Change, USA 77
National Science Board, USA 22, 32
National Science Foundation (NSF) USA 27, 72, 77, 83
National Trust 159
Natural Environmental Research Council, UK 2
NBF (new biotechnology firms) 115–23
Nelson, R. 27, 28, 30, 33–5, 41
Netherlands
  and CERN 104, 105–7
  research expenditure 23–4, 121–2
  technology 36–7
neutron beams 66
Nicholson, Sir Robin 63
Nimrod cyclotron 102
NOAA *see* National Oceanic and Atmospheric Administration
Noma, F. 30, 34
nuclear industry
  and public health 52, 149
  and public opinion 146, 149, 154, 157, 160, 164
nuclear physics 9, 14, 66, 114

OECD 22–5, 28, 113, 147
office machinery industry 133–40, 142
Office of Science and Technology Policy, USA 77–8
operational research 59
opportunity cost analysis, and management of science 9, 12, 140–1
Organisation for Economic Cooperation and Development (OECD) 22–5, 28, 113, 147
Our Changing Planet: The FY 1900 Research Plan 78
overhead costs in research 89–90

patents and intellectual property rights 91–3, 114
Pavitt, Keith x, 3, 31, 40
'peer group' evaluation of research 80, 88, 149, 153
penicillin 59
Perkin, W.H. 39
Perkins, Professor Don 89
Perrin, William 105
perspex 59
pharmaceutical industry 30, 113–14, 123–5
Phillips, Sir David 16, 89
physics, citation indices 60
Platt, J. 39
polythene 59
Price, Derek de Solla 22, 82
Priestley, Joseph 58
protein engineering 114
public funding of research 26–9, 41

'public good', and science 27, 28–9
public opinion
and research ix, xii, 14, 62
and science policy 79, 143–66
and Sellafield 149, 154, 157, 160
pure research *see* research, basic

radiation 50–1, 66
radio astronomy 33
radiological protection 50–1, 52, 63
radionuclides 50, 52
radon 52
Raman, Sir Chandrasekhara 66
Raugel, P.J. 116
research
basic 40, 49, 63–7; funding 28–9; interaction with technology 27–35; nature of institutions 37–9; planned or unplanned 34–5 source of technical skills 22, 33–4; in universities 28, 37–8
collaborative 65–6, 82, 93–5
and commercial innovation 34–5, 43
contribution to business 28–34
costs xi, 70–2, 89–90, 140–1; of instrumentation xi, 71–6 *passim*; US 70–2
developmental 86–7
diversity in 21, 40, 46, 76
dual support system of funding 55, 56, 64, 66, 67
efficiency 38
and employment 22, 25
and European Community 127, 132
evaluating 5–6, 10–13, 57–8, 63–4, 87–8, 140; by 'peer group' 80, 88, 149, 153; economic returns 29–30
expenditure ix, 60, 62, 121; and Accounting Standards Committee 67; analysed by country and field of research 121–3; in Western Europe 23–4
as 'free good' 28–9, 35, 40
funding 26–9, 41; from overseas xii, 35–7, 40, 41, 131–42; private 27–9, 40, 67; public 26–30, 31, 41–2, 51–8, 60–2, 89–95
growth in 22–5
and industry links 58–9, 94–5, 116, 119–20, 123–7, 139
international organisations ix, 3, 66, 98–111, 138
multidisciplinary approach 65
and multinational companies 35–7, 41–2, 131–42
and public opinion ix, xii, 14, 62, 143–166
strategic 40–1, 49–50, 63–5, 63–7, 86–7
tactical 49, 50–1, 63, 86–7
see also research and development; science
Research Councils, UK ix, 1, 5, 9, 43, 65–7, 92, 127
research and development, upstream and downstream 164
resources
allocation 1–2, 5–11 *passim*, 13–15, 83
sharing 65–6, 82
Rhône Poulenc 124
risk-taking 27, 114, 116–18, 128
Rochester University, USA 124
Rosenberg, N. 25, 26, 31, 35
Royal Aircraft Establishment, UK 47
Royal Society 16, 58

Salford University, UK 90
Sandoz 124
Sapienza, A. 116
'Save British Science' campaign 57, 64, 143
Scandinavia and CERN 103, 106
science
defined 86–7
as economic activity 26–30

science *(cont'd)*
  education 22, 25, 26, 32–4, 41–2, 54–5, 67; government involvement 51, 62
  education, USA 32–4, 72, 81–3
  and Greenpeace 150, 164
  historical role 58–62
  and public opinion xi, xii, 14, 62, 143–166
  skills provision 41, 72, 81–3; *see also* science, education
  and universities 52–3, 54–5
  see also research
Science and Engineering Research Council, UK 2, 5, 6, 40–1, 102
science and government *see* government and science
science policy
  accountability principle 1–2, 57, 65
  'arm's-length' principle 53
  'centres of excellence' 65
  corporate planning x–xi, 5–7, 57, 65
  dual support system 55, 56, 64, 66, 67
  and economies of scale 37–9
  and public opinion 143–66
  research evaluation 5–6, 10–13, 57–8, 63–4, 87–8, 140
  role of civil servants 17–20
  UK 1–20, 37–9, 47–67, 85–97
  USA 69–83
Science Policy Support Group (SPSG), UK 1, 148, 153
Science Research Council (SRC), UK 53
science and technology
  in business 28–9, 30–4
  interactions 28–35, 31–3
  international context 35–7, 64, 67
  linked to economic decline 62
Science Vote 2, 62
science, economics, management relationship 9–17
scientists
  accountability of 1–2, 57, 65
  arrogance of 1–2, 13–15
  as managers 9–17
  pay, UK 62–3, 122
Sellafield, UK 149, 154–60
Semenov, N. N. 48–9
Senker, J. 40
Sharp, Margaret xi–xii, 40
Skea, J. 38
skills, science
  availability 139
  cost of 139–40
  demand for 25, 33–4, 40–1
  shortages 63, 67
  training *see* science, education
  small firms' role, biotechnology 115–23
Smith, Adam 26
Social Science Research Council (SSRC), UK 53
Soete, L. 32
space programme 74
SRC *see* Science Research Council
SSRC *see* Social Science Research Council
Star Wars 152
'steady state' concept of science x, 1, 22–5
Stoneman, P. xii, 27
strategic research 40–1, 49–50, 63–5, 86–7
Strausse, Norman 16
Sumitoma Chem 124
superconductivity 74, 76, 151
Sweden
  and CERN 106
  research expenditure 23–4, 134–5
  technology 36–7
Switzerland
  research expenditure, 23–4
  technology 36–7
synchroton radiation 66

tactical research *see* research, tactical
Taggart, J. H. 139
Tagore, Rabindranath 54–5, 66
Takeda Chem 124

team management 15, 19
Techno Venture Management, W.Germany 118
technological innovations 61, 62
technology and science interactions 27–35, 31–3
terylene 59
thermodynamics 86–7
Transgène, France 117
transistors 27
transnational companies *see* multinational companies
Trend Report 53

UFC, UK 90, 96
UK
  apprenticeship schemes 62
  and biotechnology 116–18, 119–20
  and CERN 98–109
  citation indices 60, 122–3
  collaboration with USA 108, 109
  Council of Scientific Policy 60
  economies of scale in research 38
  Industrial Revolution 58–9
  Laboratory of Molecular Biology 119
  Medical Research Council 119
  research: expenditure 22–5, 27, 108–9, 121–2; funding 26–31, 35–7, 40, 41, 42, 51–8, 60–2, 89–95, 131–42; growth 22–5; usefulness 40
  Research Councils 1, 5, 9
  technology 32, 33, 36–7, 39, 61
Union Carbide 124
universities
  and industry links 58–9, 94–5, 116, 119–20, 123–7, 139
  and research costs 70–2, 89–90
  and research funding 57, 63–4, 72–5, 89–95
  and science 7–8, 52–3, 54–5, 85–97
University of Berlin 59
University Finance Committee, UK ix
University Grants Committee (UGC), UK 6, 53, 55, 57
US Geological Survey (USCS) 77
USA
  and biotechnology 115–18, 119–20
  citation indices 60, 122–3
  Committee on Global Change 77
  Department of Agriculture 77
  Department of Energy 77
  Department of Interior 77
  Environmental Protection Agency 77
  Gramm, Rudman, Hollings Act 73
  National Aeronautics and Space Administration 77
  National Oceanic and Atmospheric Administration 77
  National Science Board 22, 32
  National Science Foundation 27, 72, 77, 83
  research: economic return 29–31; economies of scale 38; expenditure 22–5, 27, 21–2; growth 26
  technology 31–7, 61
USGS *see* US Geological Survey
USSR
  and CERN 101, 106, 108
  citation indices 60
UV-B irradiance 71

venture capital and biotechnology xii, 110–19
Verry, Harold 100–3
Vickers, J. 27

Wyatt, James 58
Wedgwood, Josiah 58
Wellcome 125
West Germany *see* Germany, West
West Haven Research Center, USA 124

Willems, CERN 104
Williamson, O. 41
Wyatt, G. 38
Wynne, B. xii

Yale University, USA 33, 124

Young, S. 138
Yoxen, Edward 148, 153
Yuan, R. 118

Ziman, John 1, 11, 22